AF839646

Developing Customer Service Targets by Assessing Customer Perspectives

The mission of the Awwa Research Foundation (AwwaRF) is to advance the science of water to improve the quality of life. Funded primarily through annual subscription payments from over 1,000 utilities, consulting firms, and manufacturers in North America and abroad, AwwaRF sponsors research on all aspects of drinking water, including supply and resources, treatment, monitoring and analysis, distribution, management, and health effects.

From its headquarters in Denver, Colorado, the AwwaRF staff directs and supports the efforts of over 700 volunteers, who are the heart of the research program. These volunteers, serving on various boards and committees, use their expertise to select and monitor research studies to benefit the entire drinking water community.

Research findings are disseminated through a number of technology transfer activities, including research reports, conferences, videotape summaries, and periodicals.

Developing Customer Service Targets by Assessing Customer Perspectives

Prepared by:
Eric Rambo, **Robert Baumgartner**, and **Carrie Koenig**
PA Consulting Group
2711 Allen Boulevard, Suite 200, Middleton, WI 53562

Sponsored by:
Awwa Research Foundation
6666 West Quincy Avenue, Denver, CO 80235-3098

Published by:

Awwa Research Foundation
Advancing the Science of Water®

IWA Publishing

DISCLAIMER

This study was funded by the Awwa Research Foundation (AwwaRF). AwwaRF assumes no responsibility for the content of the research study reported in this publication or for the opinions or statements of fact expressed in the report. The mention of trade names for commercial products does not represent or imply the approval or endorsement of AwwaRF. This report is presented solely for informational purposes.

Published by IWA Publishing, Alliance House, 12 Caxton Street, London SW1H 0QS, UK

Telephone: +44 (0) 20 7654 5500; Fax: +44 (0) 20 7654 5555; Email: publications@iwap.co.uk;
Web: **www.iwapublishing.com**

AwwaRF report number 90988F. Originally published by AwwaRF for its subscribers in 2003.
IWA Publishing version published 2004.

ISBN 1 84339 8834

CONTENTS

TABLES

FIGURES

FOREWORD

The Awwa Research Foundation is a nonprofit corporation that is dedicated to the implementation of a research effort to help utilities respond to regulatory requirements and traditional high-priority concerns of the industry. The research agenda is developed through a process of consultation with subscribers and drinking water professionals. Under the umbrella of a Strategic Research Plan, the research Advisory Council prioritizes the suggested projects based upon current and future needs, applicability, and past work; the recommendations are forwarded to the Board of Trustees for final selection. The foundation also sponsors research projects through the unsolicited proposal process; the Collaborative Research, Research Applications, and Tailored Collaboration programs; and various joint research efforts with organizations such as the U.S. Environmental Protection Agency, the U.S. Bureau of Reclamation, and the Association of California Water Agencies.

This publication is a result of one of these sponsored studies, and it is hoped that its findings will be applied in communities throughout the world. The following report serves not only as a means of communicating the results of the water industry's centralized research program but also as a tool to enlist the further support of the nonmember utilities and individuals.

Projects are managed closely from their inception to the final report by the foundation's staff and large cadre of volunteers who willingly contribute their time and expertise. The foundation serves a planning and management function and awards contracts to other institutions such as water utilities, universities, and engineering firms. The funding for this research effort comes primarily from the Subscription Program, through which water utilities subscribe to the research program and make an annual payment proportionate to the volume of water they deliver and consultants and manufacturers subscribe based on their annual billings. The program offers a cost-effective and fair method for finding research in the public interest.

A broad spectrum of water supply issues is addressed by the foundation's research agenda: resources, treatment and operations, distribution and storage, water quality and analysis, toxicology, economics, and management. The ultimate purpose of the coordinated effort is to assist the water suppliers to provide the highest possible quality of water economically and reliably. The true benefits are realized when the results are implemented at the utility level, the foundation's trustees are pleased to offer this publication as a contribution toward that end.

Edmund G. Archuleta, P. E.
Chair, Board of Trustees
Awwa Research Foundation

James F. Manwaring, P.E.
Executive Director
Awwa Research Foundation

ACKNOWLEDGMENTS

The authors of this report wish to thank the following water utilities and individuals for their participation, both enthusiastic and patient, in this project:

District of Columbia Water and Sewer Authority, Washington D.C., Michelle Cowan, Jean Wilson, Donna Lewis, Shina Omokanwaye

City of Kissimmee Department of Water Resources, Kissimmee, Fla., Brian Wheeler, Barbara Arrant

Nashville Metro Water Services, Nashville, Tenn., Scott Potter, Martha Segal, Shannon Wray

Phoenix Water Services Department, Phoenix, Ariz., Michael Gritzuk, Ken Kroski, Theresa Marquez

We also thank the Project Advisory Committee (PAC) for their help and advice. Serving on the PAC were Karen Snyder, Hendersonville, Tenn.; Tony Gipe, Sarasota County Environmental Services, Sarasota, Fla.; Sharon Metz, Columbus Water Works, Columbus, OH. Special thanks to Linda Reekie, AwwaRF project manager, who oversaw the research and supported our work in so many ways.

EXECUTIVE SUMMARY

The Awwa Research Foundation (AwwaRF) has initiated an aggressive program to study customer satisfaction issues pertaining to water utilities. AwwaRF and the drinking water industry are striving to achieve the highest possible levels of customer satisfaction. To do this, utilities will need to develop standards of performance that are informed by the perspectives of customers themselves, and the performance standards that utilities strive to achieve will have to be meaningful and important to customers. Such information can be gained only by soliciting customer input.

Thirty years of customer satisfaction research have made it obvious that insight into customer perspectives is difficult, expensive, and inherently uncertain. It is important that the water industry not merely import a superficial form of customer satisfaction research from other industries. Especially to be avoided are measures of customer satisfaction that are no more than general ratings of overall performance. This overall rating approach does not measure what customers *expect* and how they evaluate *specific attributes* of customer service, both of which are essential to guiding managers on service improvements.

This report develops a research process and a set of data collection tools that produce a comprehensive view of customer service performance requirements, expectations, and satisfaction. The data this research produces are actionable and supplied in a format that makes them useful in utility managers' decision-making.

AN IMPROVED APPROACH TO CUSTOMER SATISFACTION MEASUREMENT

Satisfaction researchers generally view customer satisfaction as a comparison that people make between a particular consumption experience—whether it is a drink of water or an interaction with a customer service representative—and *expectations* about the experience that are built up from past experiences. A fulfillment response is created by the relationship between the two: experience and expectation. For instance, it is well understood in the industry that *changes* in the taste of water have a greater impact on customer satisfaction than any absolute qualitative experience. People can become habituated to taste components; but taste changes create a recurrent point of comparison.

This suggests that the best way to approach satisfaction is to pose to customers a comparison among different levels of service and estimate a profile of satisfaction with them. The typical point-estimate approach does not provide information about the level of their expectation, or about the comparison they are making.

This research describes a method for estimating satisfaction profiles for each significant way a utility comes into contact with customers. It describes a model of these different points of contact, but the method also can be adapted to the particular needs and concerns of utilities. The value of this approach is that it gives utilities information not only about where their current performance stands, but about the areas where performance improvement will have the greatest impact on satisfaction, or where economizing measures will have the smallest impact.

METHOD

Utility managers need information that helps them allocate resources to specific functional areas and to identify specific service levels that will increase satisfaction in a cost-effective manner. Knowledge about current satisfaction levels, especially when defined in broad terms, is inadequate by itself. To be useful as a management tool, customer satisfaction research should answer questions, such as:

- What customer service improvements can I make to give me the largest increase in customer satisfaction?
- Where should I *not* try to make improvements in customer service, because there will be little or no payoff in increased customer satisfaction?

To *optimize* service, information is also needed about the costs of providing different levels of service. With cost information, answers to some additional questions become feasible.

- What specific changes in customer service attributes give me the biggest improvement in customer satisfaction per dollar *cost* of service change?
- If I have a budget of $100,000, how can I allocate customer service resources to different functional areas to maximize my customer satisfaction ratings?

We have developed a management tool that will enable utilities to capture meaningful customer input and apply cost-benefit principles to decisions that affect customer service efforts. The approach entails three stages of data collection.

Qualitative

First, qualitative research is conducted with customers to better understand how they perceive customer service attributes and what levels of service they expect from their water utility. This qualitative research is critical to developing very specific definitions and measures of satisfaction for use in a questionnaire. The qualitative research ensures that customers are able to understand and interpret the questions in a consistent manner and give meaningful responses.

Quantitative

Second, a relatively intensive quantitative customer survey is implemented. In addition to ratings of customers' satisfaction with the current level of each service attribute, the survey also collects data on the satisfaction they would *expect* to receive if alternative levels of service were provided for that attribute.

Costing

Third, the utility undertakes a cost exercise that helps them identify the costs of delivering specific attributes of service. Since the attributes are developed initially from the customer's viewpoint, not the utility's, many of the costs cut across the usual organizational boundaries and budgetary categories. This is an opportunity for the utility to do some contingency planning, develop *what if* scenarios related to increasing and economizing on the various attributes of service.

This method is not merely one we recommend; it is a method we have followed ourselves. Four case studies of the method were undertaken, with the help of AwwaRF and four partner utilities:

Case Study 1. D.C. Water and Sewer Authority, Washington, D.C.
Case Study 2. Kissimmee Department of Water Resources, Kissimmee, Fla.
Case Study 3. Nashville Metro Water Services, Nashville, Tenn.
Case Study 4. City of Phoenix Water Services Department, Phoenix, Ariz.

Two focus group discussions were conducted in each service territory of each partner utility, for a total of eight group interviews. Based on the results of the focus groups a model of customer service was developed. Next, a detailed customer satisfaction survey based on this model was administered by mail. Enough residential customers were randomly selected from each service territory to expect a return of 600 surveys, for an expected total return of 2400 surveys. The same survey was also administered to a nationally representative sample large enough to expect a return of 1200 surveys. In addition to the customer satisfaction survey, personnel at each of the partner utilities completed a cost exercise survey.

FINDINGS

Qualitative: Focus Groups

The focus group research produced a number of important findings.

- All groups clearly viewed good customer service as service that makes the needs of the customer paramount. Courtesy and friendliness must be part of this attention, but good customer service also involves "being listened to."
- Another aspect of good customer service is that employees should be very knowledgeable about their work.
- The focus group participants had a clear sense that management is responsible for the treatment they get from employees and that good customer service reflects strong management.
- The taste of water was a source of general complaint. Yet this opinion about the taste of tap water did not translate into criticism of the utility in every case. Many people seem to have come to accept the water's taste, in some sense do not see it as being the utility's fault, and do not equate this taste with water that is unsafe to drink.
- In every location, there was at least one person who expressed concern over the safety of the water system from terrorists. There appeared to be general agreement that this was a concern, but it never became a dominant or prolonged theme of the conversation.
- There were mixed sentiments concerning water quality information. The idea that information in general is good was widely accepted, but only a few appeared adamant that they wanted more information about the water, possibly from an independent source.
- When asked what they expected from field service representatives who come to their house, the important attributes that were mentioned in most groups were timeliness, a

reasonable appointment window, knowledgeable representatives, and complete restoration of property when the work is done.

- Overall, participants said they were quite satisfied with the actual service they had received, both over the phone and walking in to water utility offices.
- Most participants expect the utility to play a broader role in the community than providing water and sewer services.

Quantitative: Satisfaction Survey

Overall satisfaction with the partner water utilities is moderately high (averaging 80 points over all samples). The lowest satisfaction, perhaps not surprisingly, is with water quality (average 69 across all samples) and price (average 70 across all samples).

Average satisfaction with 17 attributes of service is presented in the Figure ES.1.

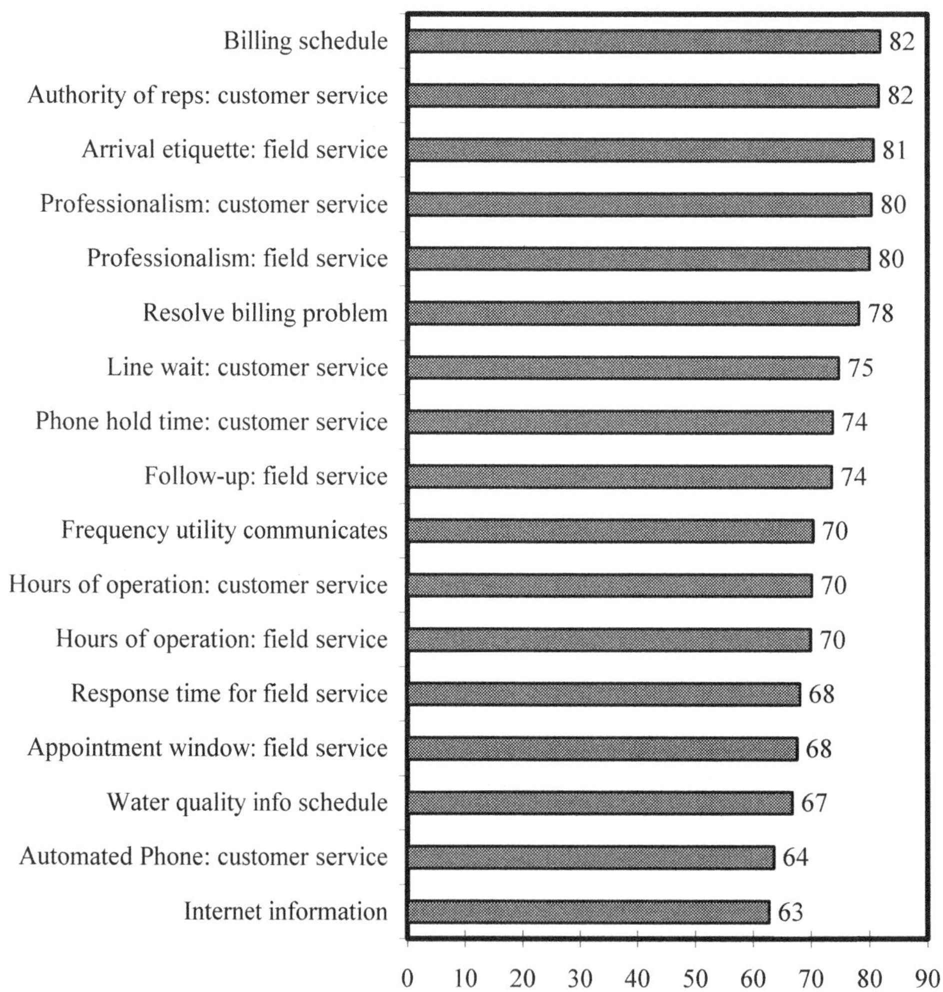

Figure ES.1 Comparison of average satisfaction among seventeen attributes of service

A central part of the method was asking survey respondents to rate how satisfied they would be at different levels of service—high, medium, and low—for the *same* attribute of service. Results of these ratings are presented in Chapter 3. Perhaps the most important finding from this part of the survey is that satisfaction ratings of the different service levels were remarkably similar across all five of the survey samples (i.e., the four local utilities plus the national survey). This suggests—though only tentatively—the existence of a fairly general set of performance standards across the country.

Cost Exercise

Cost information was gathered in a survey of the four partner utilities. Utilities were asked what measures they would take if they wanted to improve service on each of the attributes covered by the satisfaction survey. Then they were asked how much those measures would cost and what impact the measures would have on customer satisfaction. The utilities were asked, where possible, to provide two levels of service change in each direction, so cost changes for a total of five service levels would be available: current service, first improvement, second improvement, first economy measure, and second economy measure. The findings are presented in Chapter 3. These findings feed into the next stage of the research, the identification of optimal levels of service across all attributes.

Optimal Service Model

The previous findings were used to model optimal levels of service, using a linear programming approach. Linear programming is an algebraic approach to solving a set of equations to maximize (or minimize) a single value, such as satisfaction. In this case, satisfaction is defined as the sum of the satisfaction derived from each attribute of service.[1] The constraint on maximizing satisfaction is the cost of providing service at each level. The value of the cost function can be set equal to zero in order to make changes in service neutral with respect to overall cost, or set above or below zero to simulate investments and cost cutting.

The findings of the optimization modeling are presented in Chapter 3. The estimated satisfaction increase from the model's recommended service changes ranges from 1% to 17%. Each partner utility has a unique configuration of recommended changes. In general, the model shows that customer communication costs and representative training costs leading to more effective interpersonal relations have the most impact per dollar invested. A series of further analyses are also undertaken to show the robustness of the method.

RECOMMENDATIONS

Based on the method developed in this report and the case studies that were conducted, we make the following recommendations:

- Water utilities should find the resources necessary to conduct customer satisfaction research.
- Utilities should develop customer satisfaction surveys and focus groups that ask detailed questions about operations, and not the kind that ask only general questions about satisfaction with the utility.

[1] Satisfaction with each attribute is defined as the product of the percentage of customers at each level of the attribute and the weighted satisfaction derived at each level. This is defined more fully in Chapter 3.

- Utilities should develop customer satisfaction surveys that go beyond estimates of satisfaction with the current service experience, to estimate satisfaction profiles with different levels of service.
- Utilities should collect cost data that is directly linkable to customer satisfaction data.
- Utility managers should not assume that because overall ratings of customer satisfaction for their utility are high, no significant improvements can be made.
- Recommendations of an optimization model should be treated as only one important source of information, to be confirmed by other information and judgments.

CHAPTER 1
INTRODUCTION

All of the changes that constitute the so-called "new economy" have been driven by competition. In the new economy the emphasis of business shifts from goods to services, from products to customers, from individual transactions to enduring relationships, and from attracting customers to retaining them (see Rust et al. 2000, p.5). It may be true that the water industry is relatively late to experience significant competition—even now competition in many places is only a distant concern. As service improves throughout the economy, however, any industry that lags behind may be liable to attract the attention of both its customers and its potential competitors.

Although some leading utilities have already taken steps in this direction, it is only now that the water industry has reached a moment where it *must* turn a portion of its attention from technical competence to customer insight. In coming somewhat late to this moment it has the clear advantage that it can learn from others' mistakes. Thirty years of customer satisfaction research have made it obvious that insight into customer perspectives is difficult, expensive, and inherently uncertain. Many of the methods that have been developed to measure customer satisfaction have turned out to be of little value, either because they did not measure what they intended to, or because whatever they measured was of little practical value to managers. Thus, coming late to the field, the water industry has an opportunity to leap over many failed efforts and to tailor an approach from the best knowledge of a well-developed field.

AwwaRF and the drinking water industry are striving to achieve the highest possible levels of customer satisfaction. To do this, utilities will need to develop standards of performance that are driven by the perspectives of customers themselves. Such information can be gained only by soliciting customer input.

It is important that the water industry not merely import a superficial form of customer satisfaction research from other industries. Especially to be avoided are measures of customer satisfaction that are no more than general ratings of overall performance. This overall rating approach does not measure what customers expect and how they evaluate specific attributes of customer service, both of which are essential to guiding managers on service improvements.

In this report, a research process and a set of data collection tools are outlined, which produce a comprehensive view of customer service performance requirements, expectations, and satisfaction. The data this research produces are actionable and supplied in a format that makes them useful in utility managers' decision-making.

The project also goes one step beyond the normal approach to customer satisfaction research, providing a *management tool* that will help utility managers make informed decisions about the *costs* as well as the benefits of allocating resources to customer service functions. As discussed more fully below, based on the output from this study, utility managers will be able to conduct a cost-benefit analysis of specific customer service investments. The project output will combine utility data on the cost of meeting customer-identified service targets with corresponding data on customers' expectations and satisfaction with each of these service targets.

This extra effort allows managers to ask and answer such questions as

- Which of several competing investments in customer service will give the greatest return?
- If I make an additional investment in the call center to decrease the average length of time customers are on hold, how much of a return will I get in increased customer satisfaction?
- Are there specific areas of customer service where we are overspending? In other words, can we reduce performance in certain areas without decreasing customer satisfaction?

These are all questions utility managers should be able to answer based on the results of a thorough and well-designed customer satisfaction study.

PROJECT BACKGROUND

While customer satisfaction is one of the most widely studied areas of consumer research in the water utility industry and in many other industries, it is far too common for studies to be superficial or unhelpful. The problem is usually not reluctance or inability on the survey respondents' part. Customers can (and will) generate preferences for any object or expression presented to them if they are asked to do so in the context of a survey. The problem with customer satisfaction research has had much more to do with the failure of researchers to pose helpful questions.

A typical study in the drinking water industry (as well as the electric and gas industry, to name other examples) would consist of a survey that asks customers to evaluate a dozen or more general characteristics of a company—e.g., "employees' willingness to help," "courteousness," "professionalism," or the "convenience of office hours." Customers are asked to rate their water utility on a five- or seven-point scale for each of these general measures, and results are reported item by item, with analysis consisting of little more than a narrative summary.

There are a number of problems with such a method. For one, unless respondents have a specific experience to draw upon in an area of customer service, each of these generalized ratings tends to reflect a single underlying feeling about the company. There is, thus, little variation in response from one question to another and the data become redundant. Second, and more importantly, questions posed in such a general form do not provide any guidance to water utility managers about how to improve the organization's performance in one or more of these areas of customer service. These data do not provide any insight into *why* customers rate their utility as they do. Similarly, they provide little or no guidance on what specific steps a utility manager should take to improve customer satisfaction in one of these areas.

AwwaRF has funded a number of studies that have provided useful information to the water utility on customer attitudes and best practices for customer service functions. Customer attitude surveys conducted in 1989, 1993, and 1997-98 provide insights into customer beliefs about the importance of various service and communications functions. The AwwaRF project "Customer Satisfaction: Best Practices for a Continually Improving Customer Responsive Organization" (Olstein, Stanford, and Day 2001) identifies various customer service performance targets to help utilities evaluate the effectiveness of different service delivery strategies. Another AwwaRF project, "Tools and Methods to Effectively Measure Customer Perceptions"

(Colbourne et al 2002), identifies a variety of approaches to the measurement of customer attitudes toward water quality and customer service practices of water utilities.

As utilities in the electric and gas industry have discovered (and are continuing to discover), for customer satisfaction measures to provide truly useful information the measures have to address service targets that are focused, specific, and concrete. In practice, this means the customer satisfaction surveys have to go beyond general measures ("employees' willingness to help," or the "convenience of office hours"). To be really useful, customer satisfaction measures must capture more specific aspects of customer satisfaction. This will include such aspects as

- What specific attributes of water quality, billing, call center service, etc., do customers actually care about (or which ones do they care about the most)?
- How do customers expect their water utility to perform on each of these attributes (i.e., what do they require of the utility)?
- How do customers believe their utility is actually performing on each of these attributes?
- How satisfied are they with the utility's performance on these attributes?
- What types of changes in the performance level for each of these attributes would customers see as significant (or what kinds of changes in performance does a water utility have to make for customers to notice and change their level of satisfaction)?

A number of studies, including many of those cited above, have delved more deeply into the attributes of customer service, asking about the length of telephone hold times, the number of call transfers before a customer reaches someone who can help, the length of time required to schedule a house call, and the like. Even these studies, however, produce information that is not easy for managers to act upon. The research has measured satisfaction with respect to only a single level of customer service, i.e., the level that customers currently receive. These *single point estimates* of current performance do not provide decision makers with information on how *changes* in the level of service will affect customer satisfaction. If utilities know that 65 percent of their customers report being satisfied with call center performance, which they perceive as being on hold less than two minutes on average, how can a utility manager determine what level of performance will be required to produce an increase in customer satisfaction? In other words, if the average length of time on hold is reduced from 2 minutes to 90 seconds, will customers perceive this as a meaningful change (indicated by a change in the level of satisfaction) or will this be perceived as roughly the same level of performance, with no change in customer satisfaction ratings?

From a utility manager's perspective, it is crucial to know how much a given change in the level of service will produce a change in satisfaction ratings. A manager needs to know, "If I spend more to increase the level of service, will I get any payback from the additional customer service resources I have devoted to the call center?"

To maximize the usefulness of customer satisfaction data, the projected changes in satisfaction arising from changes in the level of service provided should be measured. In effect, the customer satisfaction study should produce a series of customer preference contours for each of the key attributes that influence satisfaction with an area of customer service. The preference contours will show how customer satisfaction changes as the level of service provided changes. It is true that preference contours would be less important in the context of an unlimited budget, where managers could simply give customers the highest level of service they could devise.

However, since budget constraints are a permanent feature of economic reality for water utilities, managers have to make difficult choices about allocating a finite amount of customer service dollars and ultimately about the level of service they provide for each of the key customer service areas and attributes.

The search for common standards and targets for customer service is a natural step in the emergence of a customer-focused marketing perspective within the water utility industry. Because so much work has already been done by other industries—industries that felt the pressure of competition more urgently and earlier—the water industry now has the opportunity to leap over many of the initial stages of research to identify customer service targets and incorporate customer satisfaction data into utility decision-making processes.

In particular, the water utility industry should avoid relying solely on best-practice *benchmarks* as the basis for determining customer service standards for individual water utilities. The type of benchmarks we are referring to have been established by looking at the practices of utilities rather than the preferences of customers. This type of inward-looking benchmark provides a tremendously useful guideline and frame of reference for utility managers, but it is potentially disconnected from the customer preference set. If the water utility industry is to fully incorporate their customers' requirements and expectations for service targets into the services they provide, they will need to have a full and detailed set of customer satisfaction data and preference contours.

In addition to customer preferences and expectations, the *cost* of providing a given level of service is a fundamental condition of all customer service provision. From a water utility manager's perspective, cost cannot be ignored when customer service standards and performance targets are set. When information on the cost of providing different levels of service is combined with customer preference profiles for specific service attributes, utility managers will be able to optimize their customer service performance and return on service expenditures at any level of monetary investment.

CONCEPTUALIZING CUSTOMER SATISFACTION

Like many other terms that are used in everyday conversation, the term "satisfaction" seems straightforward only until it is opened up to careful analysis. At that point it is easy to get lost in the complexity of its meaning, and never return to the practical issues of measurement and decision-making. While that has to be avoided, a kind of naiveté about nature of satisfaction, on the other hand, has produced a great deal of the low-value research that exists. Oliver (1997, p. 7) and Vavra (1995, p. 35) point out that very few of the numerous books published on the subject explain what they mean by the term *customer satisfaction.* At least a small degree of conceptual depth is needed to capture what customer satisfaction really is and avoid numerous errors of measurement.

Satisfaction Defined

Satisfaction researchers generally view customer satisfaction as a comparison people make between a particular consumption experience—whether it is a drink of water or an interaction with a customer service representative—and *expectations* about the experience that are built up from past experiences. A fulfillment response is created by the relationship between the two: experience and expectation. When people expect mediocre service, they will be satisfied if they receive mediocre service; but they will not be satisfied with poor service. Conversely, when people expect truly great service they will not be satisfied with merely good service. Richard L. Oliver (1980) was the first to develop this insight into the role of expectations in the formation of a "fulfillment response" (see also Vavra 1995, p. 42).

Oliver (1997 pp. 72-73) argues that people may simultaneously hold a variety of expectations about a single consumption experience. They may have a notion of what would be ideal, excellent, desired, deserved, needed, adequate, minimally tolerable, or intolerable. Any of these expectations may be accompanied by a judgment of the likelihood that it will occur. Comparisons also are made with different referents; that is, the current experience is compared with different alternatives. An encounter with water utility personnel may be compared to past experiences with the same water utility, with a different water utility, with a communications or energy utility, or with retail businesses.

If consumption experience is similar to what is expected, it is *assimilated* into the expectation. That is, the expectation is reinforced but otherwise unchanged. Again, if the expectation is for low-level service, and low-level service is received, there will be no nasty surprises and people will not strictly speaking be dissatisfied. We assume that a complex set of motivations prompted them to return for more low-level service.

If an experience is different than expected, either better or worse, this causes a kind of dissonance and a search for an explanation. If there is an available explanation that fits into a more complex understanding of the experience, the expectation itself will still be unchanged. For instance, a person expects efficient service but receives something less, as long as there are circumstantial cues that explain the experience as an aberration, such as an obvious shortage of staff on that day, the expectation will not alter. The possibility exists, however, that the expectation will begin to change to *accommodate* the new experience, e.g., "Service used to be efficient but now it is not." Note that the comparison in this case is to previous experiences at the same site. The expectation will also begin to accommodate new experiences if, for example, customers have experienced a contrast at other sites that seem relevant to them. It may never occur to customers to compare service at their water utility with service at Disney World, for instance, but if other utilities or comparable retail businesses begin to offer more efficient service, the expectation about service at the water utility may begin to change, and what was once seen as satisfying will now seem less so.[2]

Clearly, these observations suggest that customer satisfaction is a fluid reality. Standards change; indeed the tendency is for them to rise. Organizations that become complacent with a good service record are likely to unexpectedly fall behind. They also suggest that a better way to approach satisfaction is to pose to customers a comparison among different levels of service and estimate a profile of satisfaction with them. The typical point-estimate approach does not provide information about the level of their expectation, high, low, or medium, or about the comparison they are making. Thus, if a customer says she is satisfied, was her expectation high, medium, or low?

Aside from these general comments about the nature of satisfaction, a number of other distinctions are useful to draw. In each case the distinction reflects an evolution in thinking about the subject. First, satisfaction can be seen to derive either from the *features* of products or from the *benefits* that products convey. Second, satisfaction can be viewed as a *moment in time*, or it can be seen as an *ongoing relationship* between a customer and a business. Third, the benefits

---

[2] In focus groups that were conducted for this research project, a number of people gave the example of Disneyland or Disney World as the epitome of good customer service.

derived from products or services can be seen to include not only an obvious *core set* but also a more subtle *peripheral set* of benefits

Product Features vs. Product Benefits

In most customer satisfaction research, the focus of attention is shifting from the specific features of the product that is consumed to the experience of customers (Rust et al. 2000, p. 6). The former view concentrated on technical competence related to the product, and the process of making it excellent. More recently it has been understood that satisfaction derives not from the features of products per se, but from the benefits that product features convey to their consumers. This distinction, although subtle, more completely transfers the focus of satisfaction research onto the perspective of customers themselves. It is customers who define the benefits they expect to receive and it is customers who decide whether they have received them; it is customers, not technicians, who define excellence. It should be noted that customers might not have an easy time articulating the benefits that they seek. It is the role of satisfaction research, however, to help them.

Relevance to the Drinking Water Industry

How does this distinction between features and benefits apply to the drinking water industry? Treated water, for instance, would no longer be the principal concern; it would be the bundle of benefits that treated water conveys from the standpoint of customers. This bundle could be quite diverse and differentially distributed among the consuming public, and it might change over time. It becomes the task of the utility to understand this complex mindset of its customers and attempt to serve it.

It may at first be difficult to imagine that there is anything complex about the benefits that water conveys to customers. Aside from the more obvious material benefits, however, water almost certainly brings various intangible ones. With many other kinds of products, such as automobiles, clothing, and the like, it is easy to see that the bundle of satisfying attributes includes things such as status or identity that are not apparent in the practical benefits of the products. Treated water perhaps is not as rich in this kind of symbolic benefit as some other products. Nevertheless, it would not be too far-fetched to suggest, for example, that some people derive satisfaction from the *idea* that water is pure. Thus, anything that detracted from the idea would lower satisfaction, even if the actual purity of the water were unaffected. The point is that this intangible aspect of customer satisfaction has not been adequately studied, and so utilities have not been able to respond to it.

Consumption Episodes vs. Customer Relationships

Another development in customer satisfaction research has been the general shift in focus from single episodes of contact with a product to an accumulation of experience with it (Sharma, Niedrich, and Dobbins 1999, p. 232; Oliver 1997, pp. 16-17). The initial conception came from a model that saw satisfaction flowing immediately from a specific consumption event. The insertion of expectations into the satisfaction process, however, helped to wean market researchers and managers away from this rather naïve view. The alternative view is that an overall judgment of satisfaction emerges from the process of remembering and comparing, and it is this overall level of satisfaction with a product and company that should be of primary concern. Certainly, for some products that do not depend on repeated transactions with customers, the single episode conception of satisfaction remains valid. For most situations, however, it is quickly losing favor.

This change in perspective has been the basis of a new approach to providing customer satisfaction, known as relationship marketing. There is a growing literature advising business managers to get "intimate" with their customers (Wiersema 1996), to "manage customer relationships" (Nykamp 2001), or to build "customer equity" through long-term relationships with their customers (Rust et al. 2000).

Relevance to the Drinking Water Industry

In the water industry, there rarely has been any alternative to a long-term relationship with customers. It is not clear, however, that utility managers have always appreciated the need to *cultivate* this long-term relationship, for instance by increasing the visibility and saliency of the utility's identity. In an era where competition is increasingly likely, it would seem prudent to take a lesson from other industries and think in terms of building customer loyalty.

Core Benefits vs. Peripheral Benefits

Core benefits are the essential reasons for consuming a product, without which the buying relationship would not exist. Peripheral benefits are those that accompany core benefits but are not in themselves essential to the buying relationship (Goodman et al. 1995). Overall satisfaction is a function of satisfaction with both core and peripheral benefits. Goodman et al. (1995) show evidence to support an intriguing hypothesis: that dissatisfaction with peripheral benefits takes a double toll on overall satisfaction. It reduces overall satisfaction directly and also indirectly, because it reduces satisfaction with core benefits.

Relevance to the Drinking Water Industry

For water utilities, the core benefit a customer derives might be—and here we are speculating, of course—the availability of abundant safe tap water. A peripheral benefit might be that the water has no unpleasant taste or that the service personnel are courteous and skilled in their work or that they do not have to hold on the phone for too long when the call utility offices. So, according to this distinction between core and peripheral benefits, dissatisfaction with hold times at a telephone center would tend to directly reduce satisfaction with a utility, and also reduce satisfaction with the quality of water, which would further contribute to the overall satisfaction. The implication is that indirect benefits must not be overlooked in the overall effort to build customer satisfaction.

PLAN OF THE REPORT

In Chapter 2 of this report, we identify an approach to conducting customer satisfaction research that builds on the definition of satisfaction we have just presented and that provides the kind of useful information we suggested should be expected of such work. The approach involves three phases of data collection: a set of focus groups, a customer survey, and a cost exercise for utilities to undertake. In discussing the approaches we describe a set of four case studies we conducted to develop the method and demonstrate its value.

In Chapter 3, we present the findings of the project. There are six sets of findings:

- Qualitative results from the focus groups
- Current satisfaction with the utility from the mail survey
- Projected satisfaction with different levels of service from the mail survey
- Weights representing the relative importance of different attributes of service in determining overall satisfaction
- Results of the cost exercise identifying the incremental cost of increasing and decreasing service
- Results of a modeling process that identifies optimum levels for providing different attributes of customer service

Chapter 4 is a discussion of the findings. We discuss the critical assumptions of this model and how to fine-tune the findings by doing sensitivity testing. We also discuss different ways this approach to customer satisfaction research could be implemented, including using only selected parts of it if desired, and what information could be gained from such partial implementation.

Chapter 5 presents a set of recommendations for utilities based on what has been learned in this research. We strongly recommend undertaking the modeling process but also caution about the inherent limitations of such research.

CHAPTER 2
A METHOD FOR CONDUCTING CUSTOMER SATISFACTION RESEARCH

Although the water industry is increasingly turning to customer satisfaction research to help guide its decision making, in too many cases the findings provide too little information, or the wrong type of information to support utility management decisions. Utility managers need information that helps them allocate resources to specific functional areas and to identify specific service levels that will increase satisfaction in a cost-effective manner. Knowledge about current satisfaction levels, especially when defined in broad terms, is inadequate by itself. To be useful as a management tool, customer satisfaction research should capture several related aspects of satisfaction:

- What attributes of customer service do customers care about most?
- How do customers expect a utility to perform on each of these attributes?
- How do customers believe the utility is actually performing on each attribute?
- How satisfied are customers with the utility's performance on each attribute?
- How much would their satisfaction change as a result of different levels of performance on each attribute?

This allows utility managers to answer a variety of questions, such as:

- What customer service improvements can I make to give me the largest increase in customer satisfaction?
- Where should I *not* try to make improvements in customer service, because there will be little or no payoff in increased customer satisfaction?

To *optimize* decision making, however, additional information is needed; in particular, about the costs of providing different levels of service. With cost information, answers to some additional questions become feasible.

- What specific changes in customer service attributes give me the biggest improvement in customer satisfaction per dollar *cost* of service change?
- If I have a budget of $100,000, how can I allocate customer service resources to different functional areas to maximize my customer satisfaction ratings?

This chapter describes a method of gathering the information needed to answer such questions as these. We have developed a management tool that will enable utilities to capture meaningful customer input and apply cost-benefit principles to decisions that affect customer service efforts. The approach entails three stages of data collection.

First, qualitative research is conducted with customers to better understand how they perceive customer service attributes and what levels of service they expect from their water utility. This qualitative research is critical to developing very specific definitions and measures of satisfaction for use in a questionnaire. The qualitative research ensures that customers are able to understand and interpret the questions in a consistent manner and give meaningful responses.

Second, a relatively intensive quantitative customer survey is implemented. In addition to ratings of customers' satisfaction with the current level of each service attribute, the survey also collects data on the satisfaction they would *expect* to receive if alternative levels of service were provided for that attribute.

Third, the utility undertakes a cost exercise that helps them identify the costs of delivering specific attributes of service. Since the attributes are developed initially from the customer's viewpoint, not the utility's, many of the costs cut across the usual organizational boundaries and budgetary categories. This is an opportunity for the utility to do some contingency planning, develop some *what if* scenarios related to increasing and economizing on the various attributes of service.

This method is not merely one we recommend; it is a method we have followed ourselves. We have undertaken four case studies of the method, with the help of AwwaRF and four partner utilities:

Case Study 1. D.C. Water and Sewer Authority, Washington, D.C.
Case Study 2. Kissimmee Department of Water Services, Kissimmee, Fla.
Case Study 3. Nashville Metro Water Services, Nashville, Tenn.
Case Study 4. City of Phoenix Water Services Department, Phoenix, Ariz.

In Chapter 3, we present a detailed set of findings from the research. Thus, the current chapter both describes what was done and recommends research that other utilities could conduct.

The method is presented as a complete package. However, utilities could employ all or only some of it, depending on their needs and resources. A number of the findings reported in Chapter 3 should be directly useful to decision makers without taking any extra measures. Utilities that want to conduct their own research can follow our method exactly, or adapt its recommendations to their own needs. Chapter 4 discusses a variety of ways the methods reported here could be altered to suit different circumstances.

OVERVIEW OF THE METHOD

Two focus group discussions were conducted in each service territory of each partner utility, for a total of eight group interviews. Based on the results of the focus groups a model of customer service was developed. Next, a detailed customer satisfaction survey based on this model was administered by mail. Enough residential customers were randomly selected from each service territory to expect a return of 600 surveys, for an expected total return of 2400 surveys. We also administered the same survey to a nationally representative sample large enough to expect a return of 1200 surveys. In addition to the customer satisfaction survey, personnel at each of the partner utilities completed a cost exercise survey.

In each of the following sections, we briefly explain the logic of the different steps involved in the research. Then we return to a more detailed description of each step, describing in detail what we did, and providing basic information that will guide utilities that either want to undertake similar research on their own, or to intelligently consume services from professional research organizations.

Focus Groups

The movement to provide higher levels of customer satisfaction has been built upon a single premise: *listen to your customers*. The best way to gain initial insight into customers' own perspectives on service, and to learn their expectations for water utilities, is to have them speak about their own experiences, beliefs, and feelings. Focus groups are a key method for obtaining this insight. We used them in this research as the foundation for all subsequent work.

Focus groups are group interviews, usually involving from 6 to 10 customers and a group moderator. The moderator asks questions and keeps the conversation on track, while seeking to develop a good environment for the expression of divergent views. Typically, between two and six separate group interviews are conducted. This ensures that a variety of perspectives are heard. The purpose of focus groups is to obtain information that is rich and relatively deep, but neither broad nor generalizable to a whole population. Even so, it normally requires only a few focus groups to evoke the most common experiences and sentiments.

As focus group participants share their views, they also hear the comments of others and react, either agreeing and elaborating or expressing alternative views. Out of this give-and-take process emerges a more complex image of public attitudes than can be obtained in one-on-one interviews. We learn about the common sense that people use to understand and negotiate a particular facet of their world. They speak about what they like and do not like, and why, and they explain how they behave with respect to these preferences. A careful analysis of focus groups takes away something else that is equally important: the language used to express these preferences and activities.

The focus groups for this project were designed to help identify those aspects of service that are the most salient to customers, and the levels of service that make a meaningful difference in their evaluation of the water utility. The focus groups also helped to identify appropriate language for the subsequent survey, in order to communicate clearly and accurately with customers regarding service targets. In addition, the information taken from focus group research can be built into future public communication efforts. For a discussion of focus group methods, see Appendix A.

Modeling Service

Based on the focus group findings, and in consultation with the Project Advisory Committee (PAC) and utility partners, a model of customer service was developed. The model is an attempt to capture all possible ways a water utility affects customer satisfaction. In principal, even the most trivial service changes should be modeled, inasmuch as they may be an incremental source of improved overall satisfaction. As a practical matter, however, a degree of selectivity must be used to avoid an overly burdensome customer survey.

First, we defined a broad set of *categories* of service. These take into account the general organizational structure of the utility. For this study, the categories of service were:

- Water delivery
- Billing
- Field service
- Walk-in and call-in service
- Customer communications

Within each category of service are distinct *attributes* of customer service. The attributes are very concrete facets of customer experience that are under the control of the utility. These are the fundamental unit of the model. For the present study the attributes of service were:

- Water delivery
    - Taste
    - Odor
    - Color and clarity
    - Pressure
    - Healthfulness
- Billing
    - Bill problem resolution
    - Billing schedule
- Field service
    - Hours of operation
    - Appointment window
    - Response time
    - Arrival etiquette
    - Professionalism
    - Follow-up
- Walk-in and call-in service
    - Hours of operation
    - Authority of representatives
    - Professionalism
    - Automated phone system
    - Phone hold time
    - In-person line wait
- Customer communications
    - Frequency of utility communication
    - Water quality information schedule
    - Internet information

The modeled attributes of service are defined in detail below (see pg. 17). As discussed on page 16, water delivery attributes are not included in the optimization model. This is because the reliability of responses to a written description of these attributes, since they are so subjective, is doubtful.

Finally, for each of the above attributes a set of different *levels* of service was defined. For instance, call-in customer service can be available only during normal business hours, or during extended weekday hours, or weekday and weekend hours, up to 24/7 service. Each level of service is assumed to have a different impact on customer satisfaction. By moving service delivery up and down these levels, changes in customer satisfaction, and in costs, are achieved. The size of the effect is estimated by means of a customer satisfaction survey. The levels of service are defined in Chapter 3 (pp. 37-58).

Survey Research

While the purpose of focus groups is to obtain rich, nuanced insight into customer perspectives, surveys are intended to provide information that accurately represents the distribution of attitudes in the entire population of customers. The survey questionnaire, or *instrument*, is designed to capture information required by the customer service model. In particular, the survey questions for this research addressed:

- Overall satisfaction with the water utility
- Current satisfaction with the categories of customer service
- Current satisfaction with the attributes of customer service
- Projected satisfaction with different levels of service relative to each of the attributes.
- Information about the percentages of customers who have had a variety of experiences with the utility, such as water quality problems, interactions with field service representatives, receipt of water quality information, etc.
- Demographic information

Cost Exercise

For utility managers to make intelligent choices about improvements in customer service, they need to know the costs or cost savings involved in changing levels of service. This constraint against the improvement of customer service performance is always present. If it were not, managers would have no problem deciding which level of service to offer; it would always be the highest level imaginable. That clearly is unrealistic because cost is an important consideration.

The cost exercise is designed to estimate the cost of providing the attributes of service at levels both higher and lower than the one currently offered. With this additional information we know not only how much satisfaction is gained or lost as service moves up and down, but also how much the change will cost or save in dollars. In combination we can estimate the marginal cost of customer satisfaction.

The cost exercise is particularly challenging because the costs of providing any attribute of customer service—as defined by *customers*—are often incurred by multiple units within an organization. Budgeting practices typically do not isolate costs in a way that conforms to our model. To a considerable degree, however, we believe this exercise has value in and of itself because it forces utilities to think about their operations in a new way. Although the cost estimation exercise is necessarily more modest in depth and scope than a full-blown benchmarking exercise, it allows us to calculate a good estimate of costs per unit of customer satisfaction.

Data Analysis

For the initial analysis, standard descriptive statistics were all that was necessary. These findings identify current satisfaction with general categories of service, as well as with the specific attributes. Using the data about satisfaction with different levels of service, a customer preference profile was developed. Even without the cost analysis, utility managers could use this profile to see where marginal increases in satisfaction begin to decline with increases in service, giving them a general sense of how much investment to make in each service area.

For the service optimization modeling, satisfaction coefficients were estimated for each level of each customer service attribute. These coefficients were used in a linear programming model to calculate maximum satisfaction relative to a specified budget. We can use this analysis to determine where current money should be best spent, or to determine where added investments in service would pay the greatest dividends in customer satisfaction.

DESIGN AND IMPLEMENTATION OF THE CUSTOMER SATISFACTION SURVEY

The customer survey was designed to get a representative measurement of current satisfaction with the utility, and of projected satisfaction with different levels of customer service. The survey also tallied a number of experiences customers may have had with the utility, such as field service calls, bill problems, water quality problems, and the like. It also was used to collect demographic information. It may be possible to design a phone survey that would gather this quantity of information, but we highly recommend using a mail survey for this type of research. Well-designed mail surveys can have a response rate that is as high or higher than phone surveys. They do take longer in the field, however. Web-based research is another option that may become more attractive in the future, but at present response-bias continues to be a concern, because many people do not have access to the Internet.

The following comments are intended both to document the methods used for the four case studies, and to provide information relevant to the design and implementation of surveys for utilities that want to field one of their own. The comments do not constitute a complete set of instructions but are somewhat selective in light of our experiences working with water utility personnel. For a more thorough introduction to survey research methods, we recommend Dillman (2000) and Hayes (1998).

For the case studies, we administered surveys to a random sample of residential customers of each partner utility, in numbers sufficient to expect a return of 600 per utility. We also administered surveys to a representative sample of U.S. households in sufficient numbers to expect a return of 1200. Our response rate expectations were based on previous experience with surveys of similar topic and length. The actual response rate was below our expectation. Table 2.1 presents the response rate for each of the five survey samples.

Table 2.1
Survey response rate for five samples

	TOTAL	Nashville	Kissimmee	Phoenix	Washington D.C.	National
Starting sample	**9598**	**1500**	**1500**	**1500**	**1498**	**3600**
Ineligible	472	25	29	5	3	410
- utility employee	1	0	0	1	0	0
- renter	17	0	3	0	0	14
- business	3	0	0	0	0	3
- deceased	11	1	2	1	0	7
- incapable (health reasons)	3	2	1	0	0	0
- language barrier	3	2	1	0	0	0
- undeliverable	244	17	19	0	0	208
- moved from original address	2	1	0	0	0	1
- receives water from private well	188	2	3	3	3	177
Adjusted sample	**9126**	**1475**	**1471**	**1495**	**1495**	**3190**
Refusal	212	29	32	32	51	68
Completed survey	**2817**	**645**	**440**	**547**	**507**	**678**
Response rate*	**30.9%**	**43.7%**	**29.9%**	**36.6%**	**33.9%**	**21.3%**

* Computed as the number of completed surveys/adjusted sample

The mail survey instrument designed for this project imposes a significant burden on the respondent. It was 16 pages in length, with 43 questions, some of which have as many as five sub-questions. We estimate it required approximately 20 minutes to complete the survey. Certainly, response rate was reduced because of this respondent burden; however, we did achieve a response rate of almost 44% for the Nashville sample, so careful implementation can yield a reasonably good result. The survey instrument we used is included as an appendix (see Appendix B). Below, we discuss implementation strategies for improving response rate. Some of these strategies were not practical for the current project, but would be available to many utilities that wanted to adopt the overall method.

The survey instrument was broken down into six distinct categories of questions. This was done partly to reduce the monotony of the task, but also to help respondents focus on specific aspects of their water service. Below is a description of the different sections of the survey.

General Satisfaction

The survey begins by asking respondents to think about their water utility in general terms, then rate their overall satisfaction on a 100-point scale (with 0 meaning not at all satisfied and 100 meaning completely satisfied). Next, respondents are asked how much they pay for water service in a typical payment period. The point is not to correlate overall satisfaction with price. We do not attempt to measure price in a way that would disentangle differences in payment period and segregate water service from wastewater and other services. Our purpose is only to call to mind the price that is paid for water, so that the next question, which asks respondents to rate their *satisfaction with* the price they are currently paying for water service, is based on a clear recall of experiences. This is a fundamental rule of good survey design: make the questions as salient as possible by evoking concrete experiences.

Water Quality

In this section, respondents are asked whether in the last 12 months they have had any problems with pressure, taste, odor, or clarity of their tap water, or any concern about its healthfulness. They were then asked to rate their satisfaction, again on a 100-point scale, with the same set of items. To reiterate the purpose of asking first about experiences is to root the satisfaction items in a clear recall of the past. These questions also give us a basis for diagnosing dissatisfaction with water service.

In addition to questions about water quality, the survey also asks respondents about their water drinking habits, whether they drink tap water or bottled water and whether the household uses a water filter. By asking these questions, we are able to get a sense of the different behaviors associated with water quality satisfaction levels.

We did not build a full set of service attribute questions for water quality, even though water quality is an important, and many think the single *most* important factor in determining overall customer satisfaction. (Our evidence does not support this contention, however.) One problem with modeling water quality satisfaction is that it is difficult to develop meaningful questions about levels of service that can be articulated in a survey. The frequency with which problems occur can be asked easily, but not questions about qualitative differences in taste, odor, clarity and pressure; and these are the crux of the matter. The second problem with approaching water quality problems with this method is the difficulty in assigning costs to changes in service levels. Note that we do not say water quality is absolutely off limits to our method. However, we made a decision, in agreement with the PAC and partner utilities, to focus our energies on other problems that are more within the traditional purview of customer service.

The Modeled Categories of Service

The categories of service to be included in the optimization modeling are all structured in essentially the same way. First, we ask a series of questions to tally particular experiences respondents have had with the utility, questions that bring past experiences to mind and also provide a basis for segmenting the responses. Next, we ask a series of questions about the modeled attributes of service, and finally we ask a question about overall satisfaction with the category of service.

The service attribute questions also have the same general form. Respondents are asked how satisfied they *would be* with three or four hypothetical scenarios. Then, they are asked to identify the scenario that is closest to their actual experience. The scenarios are presented in what is presumed to be an order of decreasing satisfaction but it is not necessary to the accuracy of results that the presumption about order is correct. If the respondents have had no experience with an aspect of service, they indicate this and respond with the level of service they would expect to receive from the utility.

At the end of each set of attributes, we ask an overall satisfaction question. Answers to these category-level satisfaction questions are used to estimate the relative importance of the attributes of service, and to develop a stable, richly informed overall assessment of the utility.

The attributes we modeled are summarized below.

Billing

- *Bill problem resolution*—If a problem is discovered on the bill, is it resolved quickly and easily, or with numerous contacts with the utility and difficult negotiations? (3 levels of service)
- *Billing schedule*—How often is the bill sent? (4 levels of service)

Field Service

- *Hours of operation: field service*—During what hours can customers obtain non-emergency field service? (4 levels of service)
- *Appointment window: field service*—When arrangements are made to send a field representative to someone's home, what is the time period within which the representative is scheduled to arrive? (3 levels of service)
- *Response time: field service*—What is the time period that elapses between a call for non-emergency service and the arrival of field service personnel? (4 levels of service)
- *Arrival etiquette: field service*—Does the field representative arrive on time, and is the customer kept informed of changes in scheduling? (4 levels of service)
- *Professionalism: field service*—Do field representatives listen to the customer, explain what they are doing, and treat the customer courteously? (4 levels of service)
- *Follow-up: field service*—Does the utility make contact with customers after service is done, to be sure the work is satisfactory? (4 levels of service)

Call-in and Walk-in Service

- *Hours of operation: customer service*—During what hours are customer service representatives available to serve customers? (4 levels of service)
- *Authority of representatives: customer service*—Do customer service representatives have the authority and the knowledge to serve customers well? (3 levels of service)
- *Professionalism: customer service*—Do customer service representatives listen to the customer, explain what they are doing, and treat the customer courteously? (4 levels of service)
- *Automated phone: customer service*—How easy is it to navigate the utility's automated phone system? (3 levels of service)
- *Phone hold time: customer service*—When calling the utility, how long do customers have to wait on hold before they can speak to a customer service representative?
- *In-person line wait: customer service*—When coming in person to the utility offices, how long do customers have to wait in line before they are served? (4 levels of service)

Utility Communications

- *Frequency utility communicates*—How often does the water utility communicate with the public about its operations and issues of special interest to customers? (4 levels of service)
- *Water quality information schedule*—How does the water utility distribute non-emergency information about water quality? (4 levels of service)
- *Internet information*—What information about the utility and the customer's account is available over the Internet? (4 levels of service)

Demographic Information

The last part of the survey asks respondents to record some background information about their household. These questions are used to determine whether systematic differences in satisfaction exist among different demographic groups. We collected information about the type of home, the age of the home, whether the respondent rents or owns the home, the size of the household, the age of the respondent, the education of the respondent, and the household income.

Survey Design

We believe the instrument designed for this research pushes the limits of what can be done in an unsupervised pencil-and-paper task and still receive quality data. In general, the rules for good survey design are as follows. Questions must be easy to understand. They must ask a specific, direct question and be worded in a non-leading manner. The meaning of a question must be explicit and avoid the tremendous capacity of humans to read divergent meanings into it. Questions should be designed so that respondents of all ages are able to read and understand what is being asked of them.

A key rule of survey design, occasionally skirted in this research, is that questions should address only a single topic. They should not be compound questions where an answer is ambiguous as to which question is being answered. In our survey, the levels of the attributes respondents evaluate sometimes combine different experiences into a single rating. For instance, the attribute Professionalism in the categories Field Work and Call-in and Walk-in Service includes the elements of listening to the customer, explaining information to the customer, and being polite and friendly. We judged it would be problematic to articulate different levels of each of these aspects of professional comportment so that the words used to evoke levels would have consistent meaning across a diverse population of customers. Also, this compounding of attributes would greatly increase the length of an already long survey. By systematically varying the presence or absence of good service on each of these elements we can get insight into the relative contribution of each to satisfaction with the attribute. We have no doubt that the separate elements of the compound attributes could be disentangled using methods that focused more specifically on them.

After each question, there are instructions to the respondent letting them know exactly what they are supposed to do. Examples of these instructions include: "Circle one answer," "Write in a number," "Check this box." The answer choices for each question should be mutually exclusive to avoid double counting certain responses. We used color on this survey to indicate to respondents where they should answer. The answer choices for each question were shaded in blue in an attempt to help respondents fill out the survey completely. We have also found that the use of color improves the response rate on surveys by as much as 5%.

Our experience has shown it is possible to design a questionnaire containing up to twenty-five attributes of customer service, with an average of four levels of service on each attribute, and still have questionnaire completion be a reasonable task for respondents. Although it would considerably increase the difficulty of implementation, it would be possible to reduce the length of the response task by splitting the set of attributes into two parts, and randomly administering only half of the set to two different samples. The general satisfaction questions and the demographic questions would be included on both versions of the survey, of course. This should increase the response rate for the survey.

Survey Implementation

The survey of utility customers consisted of up to three contacts: an initial mailing of the survey, a thank you/reminder postcard, and a follow-up survey mailing. The mail survey instrument was a 16-page booklet with a colored-paper cover. All booklets were identical except for the cover, which featured a utility logo if sent on behalf of the partner utility, or the AwwaRF logo if sent to the national sample.

The inclusion of a logo on the cover, and on the outgoing envelope in which the survey packet is contained, is vital. It establishes the legitimacy and importance of the mailing. To a large degree, we attribute the low response rate from the national sample (21.3%) to the fact that the AwwaRF logo was largely unfamiliar to respondents and unlikely to provide much intrinsic incentive to open the envelope, much less to complete the task. Governmental and quasi-governmental entities have an advantage in this regard because people generally trust that the mailing is not a commercial "come-on" and that the purpose is serious. There are, as our research indicates, various local and regional differences in the power of this influence. Utilities that sought to replicate this work in their own territory would not be saddled with the handicap of sponsor anonymity, and thus could expect the more robust response rates that the partner utilities received.[3]

The mail survey consisted of up to three contacts:

- *An initial mail survey package.* All sampled customers were mailed a survey booklet. This mailing consisted of a cover letter from the participating utility explaining the survey booklet, a mail survey booklet, and a postage-paid return envelope. The first mailing was sent on May 13, 2002.
- *A thank you/reminder postcard.* All customers were mailed a postcard within one week of the initial survey mailing (on May 20, 2002). The postcard thanked those who

---

[3] The response rate from the Kissimmee sample (29.9%) was also disappointing. We believe this response was low because of the percentage of Kissimmee residents who live in the city only during the winter months. The survey was administered in late spring and summer, and although it was sent via first class mail and would have been forwarded to summer residences in many cases, this imposed a significant time lag. We also speculate that winter residents may generally have a weaker commitment to the community than year-round residents, or at least the commitment is more remote when they are not present in the city, and thus are less likely to take the effort to provide feedback. This problem could be minimized by administering surveys during winter months.

responded and more importantly reminded those who have not responded to please do so.

- *A follow-up survey mailing.* Approximately three weeks after the thank you/reminder postcard, all nonrespondents were sent a follow-up survey mailing. This mailing consisted of a cover letter from the participating utility, another copy of the mail survey booklet, and a postage-paid return envelope. This mailing was conducted on June 21, 2002.

A number of additional measures could be taken to improve the response rate, although they add cost to the research. An initial letter could be sent prior to the first survey mailing, announcing that a survey will be taking place and asking the recipient to participate. This increases the likelihood that the recipient will notice the survey when it comes, a major hurdle in obtaining compliance, and also enhances the legitimacy of the task.

A token monetary incentive of one or two dollars could be included in the survey package.[4] The incentive does not *buy* compliance with the task in any direct sense, but it does establish a symbolic relationship of reciprocity with the survey recipient. A two-dollar incentive can increase the response rate by approximately five to eight percent. Our experience with utilities, however, suggests that many are loathe to send customers money in the mail, no matter how small the amount or how effective in improving research results.

Some researchers recommend increasing the response rate by sending the second survey mailing, or even a third mailing, by an alternative and more highly visible form of post. For instance, because we were concerned by the low response rate from the national sample after the first mailing, we sent half of the second mailing by priority mail. The intent was to increase the visibility and importance of the survey. In this case, this (expensive) strategy did not yield a higher response to the second mailing.

As a final strategy for increasing response rate, it would be possible to phone customers and inquire whether the survey had been received, whether there were any problems with the survey, whether the recipient intended to complete it, or whether a new copy should be sent. This method can be quite effective, but is often too expensive to fit within the research budget.

CONDUCTING A COST EXERCISE

The cost exercise is an internal survey of the utility, designed to estimate the cost of raising and lowering customer service levels. This information has value in and of itself, but that value is greatly increased when it is brought together with information about satisfaction changes as service levels rise and fall. It provides the constraint against which decisions about service offerings must be made.

In many ways, the cost exercise portion of a customer satisfaction study is the most difficult part of the method to implement. Since there is, in effect, only one respondent—i.e., the utility itself—the response rate has to be 100%. This requires cooperation on the part of a large number of people who gather information from a variety of departments within the business. Thus it is essential that someone influential take responsibility for completing the exercise. Since the point of the cost exercise is to recommend realistic service increases and economy measures,

[4] PA Government Services Inc. personnel have conducted research to determine the most cost-effective incentive amount to include in mail survey research (see Baumgartner and Heberlein 1984, Heberlein and Baumgartner 1978). We believe a two-dollar incentive yields the best increase in response per dollar cost.

it is also essential that each person providing information has a broad perspective on the department for which he or she is responding. It is helpful if the organizational interests of those responding are not so narrow that they are reluctant to suggest economizing measures.

For the case studies, most utility partners assigned a single person to assemble the information. We do not have good information about the degree of cooperation these important data gatherers required, or received, from others within their utilities. Nevertheless, we believe the exercise provides the best information when it is used as an occasion for departments to collectively reflect on and discuss their provision of customer service. We see it as an opportunity to plan for change, although it is essential that participants in the exercise have a clear understanding of the purpose of the information gathered and not develop unrealistic expectations about the outcome.

The Cost Exercise Worksheets

We developed worksheets for the cost exercise that are essentially survey instruments the utility administers to itself. These worksheets evolved considerably over the course of the research. Primarily, the changes were designed to simplify the task and make the information gathered more accurate. An example of the cost exercise worksheets is presented in Appendix E.

The approach we had taken in previous applications of this method was to ask utilities to estimate changes in service required to raise (or lower) the *average level of satisfaction* among customers *to one of the levels represented in the survey*. The problem with this task is that it is not a natural way to think about service provision. For instance, for the attribute Bill Problem Resolution, it would be difficult to accurately estimate what it would cost a utility to move the average satisfaction of customers from level 1.8, where the bulk of respondents say their problems are resolved *"after two or three contacts with the utility and minor negotiations"* up to level 1 where virtually all respondents would say their problems are resolved *"after one contact with the utility and with no negotiations."*[5] The factors contributing to an experience of satisfaction are only partly under the control of utilities, and not subject to the minute manipulation that is implied by this approach. For that reason we developed a new approach, which conforms to a more natural way of thinking about service.

We begin by asking the utility to evaluate what measures it would take first if it wanted to improve service or economize on a given attribute. Then, if there are other, second-tier measures the utility could take, we ask that those be recorded as well. Third, we ask how much these measures would cost to implement. Finally, the utility rates the impact the measure would have on customer satisfaction, on a seven-point scale where "1" means the impact would be "small (moves about 1/10 of customers to a lower level of satisfaction)" and "7" means "complete (moves about 9/10 of customers to a lower level of satisfaction)".

For the case studies, we provided preliminary results from the satisfaction survey to guide partner utilities about where they currently stood in terms of satisfaction with the attributes. Indeed, we delayed distributing the worksheets so that we could assemble a dataset based on

---

[5] Readers may notice a related problem with this approach, i.e., to move *average* performance to either the highest or the lowest level would require *100% of customers* to report that level of satisfaction. For middle service levels, the averages can be produced from a distribution among the possible service experiences. This is a more realistic expectation because it is rare for customers to achieve perfect consensus, even over basic issues of fact. For example, customers routinely report levels of service for billing frequency that are different than what they actually receive.

partial survey returns, in order to provide this information. Although the information was interesting to the utilities, it became less necessary to the task once we had altered the format of the questionnaire so that hitting a particular customer satisfaction mark was not necessary.

Another change we made in the midst of administering the cost exercise—after all the utilities except Phoenix had completed the task—was that we dropped a request for the current costs of providing customer service for each of the attributes. It had never been the case that this information was going to be used directly in the modeling procedure, because the relevant costs in that analysis are cost increments above and below current costs, which are placed at $0. The purpose of assembling actual current costs had been only to cause utilities to figure their expenses in a new way that was oriented by customer perspectives and service attributes rather than organizational structure. This probably is a part of the exercise that has some value, but it became clear that the worksheets were requiring a fairly substantial time commitment and we realized that streamlining the process was in order.

DATA ANALYSIS

In large measure, the analysis of the data is straightforward. An average is calculated for projected satisfaction at each level of each attribute. These are used to estimate a preference contour for each attribute. Current satisfaction with each attribute is calculated as a weighted average: a sum across the average satisfaction with each level multiplied by the percentage of customers who report that level as the service they have received. Utility estimates determine the rate at which customers notice changes in service and move to higher or lower levels of perceived service, and the preference contours determine the rate at which their satisfaction increases or decreases as they shift levels.

From this information an **m x n** matrix is created, where **m** represents the attributes of service and **n** represents up to five different levels of service—current service, two levels of increase, and two levels of reduction—and the cells of the matrix are populated with the satisfaction derived from the attribute at that level. The satisfaction values in this matrix must be weighted in two different ways, according to the *importance* of the attribute in the production of overall satisfaction, and according to the *exposure* of customers to the attribute. The importance weights used in this analysis are correlations between each attribute and a composite overall satisfaction scale. The exposure weights reflect the percentage of customers who are exposed to a particular attribute of service, estimated from responses to the customer survey.

The optimal level of customer service for each attribute is obtained by calculating the cost per unit satisfaction of each level of each attribute and reassigning dollars and people to levels of service in order to maximize satisfaction within a budget constraint. It might be possible to come close to an optimum by trial and error. That is the approach that many of the so-called simulators take. The user enters different combinations of service levels into a computer program, complete with attractive graphics, and receives an overall prediction of satisfaction for each particular combination. It is generally up to the user to discover which combinations improve or degrade the outcome. This trial-and-error approach is not the most straightforward way to find the optimum outcome, however. For the current model there are 28,625 combinations of attributes that might have to be tried. Instead, for this research we have treated the optimal solution as a linear programming problem (see Chvatal 1983, Gass 1970).

There is more than one way to model an optimal solution using linear programming. For this research, we have used the following model. The objective function (that is the function to be maximized, in this case defining overall satisfaction) is the sum of the product of the percentage of customers at each level of each attribute and the weighted satisfaction derived at each level. The function can be represented:

$$\begin{aligned}
&\text{Max}\ \ (p_{1,1}s_{1,1} + p_{1,2}s_{1,2} + p_{1,3}s_{1,3} + p_{1,4}s_{1,4} + p_{1,5}s_{1,5}) \\
&+ (p_{2,1}s_{2,1} + p_{2,2}s_{2,2} + p_{2,3}s_{2,3} + p_{2,4}s_{2,4} + p_{2,5}s_{2,5}) \\
&+ \dots \\
&+ (p_{17,1}s_{17,1} + p_{17,2}s_{17,2} + p_{17,3}s_{17,3} + p_{17,4}s_{17,4} + p_{17,5}s_{17,5})
\end{aligned}$$

where $p_{m,n}$ = the percentage of customers of the mth attribute at level n.
$s_{m,n}$ = the weighted satisfaction derived from the mth attribute at level n.

The constraint functions are as follows:

$$\begin{aligned}
&p_{1,1} + p_{1,2} + p_{1,3} + p_{1,4} + p_{1,5} = 1 \\
&p_{2,1} + p_{2,2} + p_{2,3} + p_{2,4} + p_{2,5} = 1 \\
&\dots \\
&p_{17,1} + p_{17,2} + p_{17,3} + p_{17,4} + p_{17,5} = 1 \\
&(c_{1,1} + c_{1,2} + c_{1,3} + c_{1,4} + c_{1,5}) + (c_{2,1} + c_{2,2} + c_{2,3} + c_{2,4} + c_{2,5}) \\
&+ \dots + (c_{17,1} + c_{17,2} + c_{17,3} + c_{17,4} + c_{17,5}) \leq 0
\end{aligned}$$

where $c_{m,n}$ = the incremental cost of providing service for the mth attribute at level n.

We set the value of the cost constraint function less than or equal to zero in order to make changes in service neutral with respect to cost. In other words, the solution maximizes satisfaction at current costs. The value of the cost function could also be set higher or lower to indicate where additional funds should be added or where cost reductions should be undertaken.

CHAPTER 3
THE FINDINGS

INTRODUCTION

In most research reports, the discussion of methods lends authority and credence to the report of findings. Here that relationship is somewhat reversed. The judgment that we have produced interesting findings for the four partner utilities lends authority and credence to the methods we have used, demonstrating their value for research at other utilities.

We offer six sets of findings in the following areas. Each finding is summarized below and discussed in more detail later is this chapter.

- Focus groups
- Satisfaction with current service
- Satisfaction at different levels of service
- Relative importance of the attributes for overall satisfaction
- Customer service costs
- Optimal levels of service

The first set of findings is taken from the focus group research, which formed the basis for everything that followed. These results root the broader survey research in customers' own perspectives and, although they are essentially not quantifiable, these findings in some sense come closest to the actual experiences of customers.

The second set of findings is a satisfaction rating of current services. This kind of data is the standard product of customer satisfaction research. It provides an interesting snapshot of satisfaction, especially when comparisons between utilities are possible. This kind of information, however, provides insufficient guidance to managers about what steps to take to remedy problems.

The third findings are the estimated or *projected* satisfaction that customers would derive at different levels of service from their utility. By collecting satisfaction scores for different levels of the same attributes of service, we are able to estimate a profile of preferences, which is the foundation of our cost/benefit trade-off analysis. With this profile, we can estimate the rate at which satisfaction goes up and down as service changes.

The fourth set of findings is a set of weights we apply to the attributes of service, which represent their relative importance in producing overall satisfaction.

The fifth set of findings comes from the cost exercise. The costs of providing service represent the constraint a utility works within in trying to optimize customer satisfaction. Our cost data comes from a survey we administered to the partner utilities. Here, we report the incremental cost of moving the level of service up or down, with current costs set to a value of zero.

Finally, we present the results of our optimization modeling, estimating for each of the partner utilities what kinds of customer service changes would maximize customer satisfaction, within a number of different cost constraints.

QUALITATIVE RESEARCH: FOCUS GROUP FINDINGS

The focus groups provide direct access to customer perceptions of service. We conducted a total of eight focus groups, two in each of the participating utility's service area: Nashville, Tenn. on January 8, 2002, Washington D.C. on January 10, 2002, Phoenix, Ariz. on January 15, 2002, and Kissimmee, Fla. on January 23, 2002. Between nine and thirteen people participated in each group. The participants were homeowners who received water service from the partner utility, and were selected to represent a cross-section of ages and incomes and a balance of men and women. The topic of the focus group discussion was customer satisfaction in general, and satisfaction with the participants' water utility in particular. The discussion also included topics such as water quality issues, water quality information, field service, call-in and walk-in customer service, billing, and public relations.

At three of the four locations where we conducted focus groups, the customer service personnel who observed had never before witnessed interviews on the topic of their water utility. For them, it was a startling finding that their customers have many mistaken or distorted notions about the operation of the water utility. We consider this a *dose of reality* and that a well-designed series of focus groups can be of significant value in itself.

The General View of Good Customer Service

All groups clearly viewed good customer service as service that makes the needs of the customer paramount. The sentiment was expressed numerous times that customer service representatives should make it evident that they are fully focused on the needs of the individual they are serving. For the time they are serving a customer, they should devote all of their attention to him or her. Courtesy and friendliness must be part of this attention, but good customer service also involves "being listened to." Participants gave examples of customer service employees in a variety of settings who come to a premature conclusion about what the customer needs—i.e., "What they *think* they need"— without really listening to the customer's particular request.

Style

The style of communication customers receive from customer service personnel also defines good customer service. Conveying attentiveness is essential. Participants in several locations mentioned eye contact as an important sign of attentiveness. One person articulated the point, saying she often receives set answers. "It's like talking to a recorder."

This complaint that customers are not listened to seems to reflect, in part, frustration when various demands on a company for service cannot be met because of company policy. Yet, refusal cannot be the only source of this complaint, because at least one participant reported he had received good service from a company (not the utility) precisely because an employee very politely and sympathetically informed him he could not provide a restitution that he (the customer) had requested.

Formality

Employee demeanor clearly is crucial in the customer satisfaction process. A number of participants in Nashville and D.C. mentioned over-familiarity on the part of employees as something that offends them—in particular, calling customers by their first name.

The focus group participants had a clear sense that management is responsible for the treatment they get from employees and that good customer service reflects strong management. In other words, unpleasant experiences clearly reflect negatively on the company as a whole and not simply on the employee. Customer service personnel should act professionally and work efficiently when waiting on a customer. When this occurs, participants said they are left with a positive experience that reflects back on the company as a whole.

Knowledge

Another aspect of good customer service is that employees should be very knowledgeable about their work. When service representatives are able to answer customer questions, customers are left with the sense that the staff is properly trained and the company is working efficiently. For the most part, however, the focus group participants agreed that not knowing how to handle a problem, especially if a service employee is new to a position, can be forgiven. What is not forgiven is the perception that employees are trying to hide what they do not know. Customers expect responsiveness and if the service employee cannot answer a question, customers want the employee to find the answer or transfer them to someone who can provide the answer.

Water Quality Issues

As the discussion turned from customer service in general to water utilities in particular, the topic of water quality emerged. The taste of water was a source of general complaint, though not a *universal* complaint, in every location. A majority of people in every city but Nashville said they either filter their water at home or drink bottled water. Yet this opinion about the taste of tap water did not translate into criticism of the utility in every case. Many people seem to have come to accept the water's taste, in some sense do not see it as being the utility's fault, and do not equate this taste with water that is unsafe to drink. No one in these groups expressed the view that purchasing water for taste was a sign that the utility is not doing a good job, and no one complained about the expense of purchasing bottled water. Others, however, did make the connection between water taste and water safety. These tended to be people who were more generally critical of the utility. The location with the most concern about the safety of its water was D.C., where there had been an incident five years before in which people were warned by the utility (a different entity than WASA) to boil their water prior to drinking. Several people reported they still boil all of their water, even for bathing.

One issue was more or less unique to D.C., but it bears telling. In D.C., there were numerous stories of unrepaired water breaks that appeared to play a role in the higher level of dissatisfaction with the utility. There was widespread understanding that undetected water leaks on one's own property can lead to extremely large bills. Even people who had not experienced this themselves were very aware of the possibility and, seemed concerned about it. In effect, they would be held responsible for whatever water they used, even if it was a result of factors they felt were beyond their control. Yet this "holding to customer account" appears to them to be at radical odds with the way the utility treats itself, given the stories of large breaks in their own system that run for days or weeks. There was one explicit comment of resentment that the cost of unrepaired breaks would be passed on to customers.

In every location, there was at least one person who expressed concern over the safety of the water system from terrorists. There appeared to be general agreement that this was a concern, but it never became a dominant or prolonged theme of the conversation.

Customer Communication

There were mixed sentiments concerning water quality information. The idea that information *in general* is good was widely accepted, but only a few appeared adamant that they wanted more information about the water, possibly from an independent source. In every location except Phoenix, at least one participant expressed some skepticism about the accuracy or veracity of water quality information that comes from the utility itself. This appeared to come from a general mistrust of large organizations. Most people in D.C., where the criticism of the utility was strongest, said they wanted more information. These people are also concerned about water quality in the future. They were worried that WASA will not be able to meet future demand and therefore water quality might decline. Where people generally trust the water is safe, however, they acknowledge they probably would not pay very close attention to more information.

Very few participants in any of the groups remembered receiving a water quality report. Virtually no one could mention any of its contents, except one or two people who remembered the report said the water was safe. Where the water quality report is sent along with the bill, in Nashville and Phoenix, there was agreement among the participants that putting the water quality report into the bill made it unlikely they would see it. As one person put it, "They could be putting twenty dollar bills in there and I wouldn't know it." Ironically, where the report is sent in a separate mailing, in D.C. and Kissimmee, people felt that including it in the bill would make it more likely to be seen. In D.C., the water quality report included a comment card that customers appreciated. One person had the idea that a special sticker on the bill announcing the inclusion of the water quality report would help people notice it. There was a mild sentiment that water quality information was too technical and thus difficult to read, but information distributed more often than once each year would be appreciated.

Field Service

When asked what they expected from field service representatives who come to their house, the important attributes that were mentioned in most groups were timeliness, a reasonable appointment window, knowledgeable field reps, and complete restoration of property when the work is done. Participants wanted an explanation of the work that is to be done and they did not want to be "talked down to."

A general concern with field service was the length of the appointment window that customers are given. With a few exceptions, participants in all groups agreed they did not like to set aside as much as four hours waiting for service. There seemed to be general agreement that a two-hour window was acceptable. There was also agreement that field service employees should phone if they knew they were going to be late. These issues were typically expressed as general observations, not as criticisms of the water utility. In fact, Nashville and Kissimmee participants who had received field service were complimentary about the efficiency of the service they had received.

In general, the response time of field representatives was a significant subject of complaint. The participants viewed long waits as a symptom that the business does not care about them. It is explicitly associated with cost decisions; that is, the business is cutting costs by reducing resources devoted to prompt customer service. That said, we did not record significant complaints specifically about water utilities in this regard.

Call-in and Walk-in Service

The characteristics of good call-in or walk-in service were articulated in the opening discussions about customer service. Overall, participants said they were quite satisfied with the actual service they had received, both over the phone and walking in to water utility offices. A few people in Nashville and D.C. and one in Kissimmee, complained about a long wait when they had gone in to pay a bill, but such comments typically were followed by someone who said they had the opposite experience, very excellent and considerate service.

In each location, there were expressions of frustration with automated phone systems that require numerous entries to reach an appropriate response. One person approved of a customer service situation where a representative gave out her extension number so she could be reached in subsequent calls and the automated system could be bypassed. Another person approved of a situation in which a customer service representative asked for the customer's phone number, in case the call became disconnected. The focus group participants clearly preferred to speak to a live person, and the quicker they reach a live person the more satisfied they were.

Billing

In no location was the bill considered difficult to read. In Phoenix, where a number of municipal charges are carried on the water bill, there did appear to be some confusion about which city entity receives the fees. In Kissimmee, where meters are read and bills collected by a municipal authority other than the utility, there was an almost complete lack of awareness of the utility's identity.

Community Involvement

Most participants expect the utility to play a broader role in the community than providing water and sewer services. In D.C. in particular, a number of people thought WASA should play an educational role, for instance sponsoring school programs to educate the young about water issues. By going into public schools to educate children, the children would in turn bring information home to their parents, who would also benefit from the information. Participants in D.C. remembered and appreciated efforts by WASA to help them conserve water. The principal program they remembered was the distribution of a kit to check for leaks. Participants in other locations also thought the utility should help educate the public, especially children, about water quality and conservation issues, but the sentiment was not voiced as strongly. In fact, in Nashville a number of people said that they did not want the utility to do anything except provide water.

Service Levels

Although not a finding in the same sense as those mentioned above, the focus groups were instrumental in refining and validating the levels of service that would be used in the mail survey. We asked each group to identify levels of service they would find excellent, good, and poor, for a number of attributes we anticipated using. The point of this exercise was not to tally responses but rather to look at the terms used and the categories offered to understand the range of service levels that made sense to the participants. Time constraints made this discussion less extensive than we had anticipated. Nevertheless, the focus groups helped ensure that survey recipients would be able to understand and interpret the questions in a consistent manner and give meaningful responses.

QUANTITATIVE RESEARCH: SATISFACTION WITH CURRENT SERVICE

Overall Satisfaction

In customer satisfaction research, it is common practice to ask customers how satisfied they are with general categories of service such as billing, field service, or call-in service. One problem with this approach is that customers may not have a clear or consistent view of what they should be thinking of when they consider these broad categories. They need help understanding what falls within a particular category and to stimulate recall of experiences that would be relevant.

The approach we have taken is to ask about overall satisfaction with a category only after we have explored more specific, concrete experiences that fall within that category. Hence, we ask about overall satisfaction with field service after asking about hours of service, response time, professionalism, and such. The pattern is to move from the concrete and specific to the general.[6]

The one exception to this rule is for overall satisfaction with the utility. In this case, we have opted to ask the most general satisfaction question at the beginning of the survey, in order to get impressions that have not been influenced by the survey instrument itself.[7]

In addition to overall satisfaction and satisfaction with the four categories of service—Billing, Field Service, Call-in and Walk-in Service, and Customer Communications—we asked about satisfaction with the price of water and water quality.[8] The findings are presented in Figure 3.1.

---

[6] For focus groups, the pattern is reversed. We begin with general questions to see what specific topics they will evoke.

[7] In an earlier study with Columbus (GA) Water Works, we placed an overall satisfaction question at the beginning and at the end of the survey. The rating of overall satisfaction increased by a statistically significant amount between the first and the second rating. This suggests that the process of thinking about this particular utility tended to increase satisfaction. We cannot be sure that would always be the case, of course.

[8] We actually asked five separate questions about satisfaction with water quality. We asked about pressure, taste, odor, color and clarity, and healthfulness. We did not ask an overall satisfaction question, and here present an average of the ratings on the five separate dimensions.

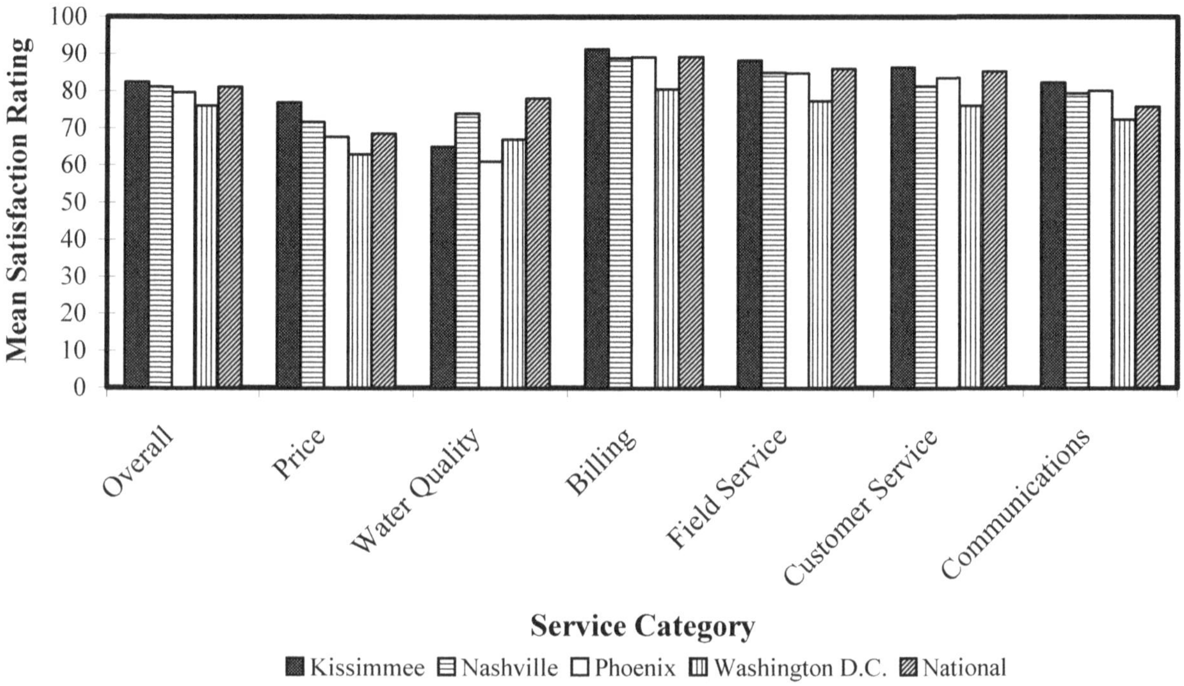

Figure 3.1 Mean customer satisfaction with water utility on general categories of service

Overall satisfaction is moderately high, averaging 80 points over all samples. The lowest satisfaction, perhaps not surprisingly, is with water quality (average 69 across all samples) and price (average 70 across all samples). Kissimmee ranks highest in all general categories of satisfaction except water quality. D.C. ranks lowest in all general categories of satisfaction except water quality. The reasons for these differences between utilities are no doubt manifold, and it is not the purpose of this research to delve into these questions. For answers we would want to compare the age of infrastructure, demographics and culture of the customer base, recent history of relations between the utility and customers, as well as utility policies and practices. Certainly the size of the service territory, the number of customers, and the location within the United States would be important factors. This is a comparison of very dissimilar entities. That is a virtue, however, for a case study project, such as this. There can be no doubt, however, that customer service benchmarking *from the customers' point of view* is a project well worth undertaking.

The problem with this information is that it does not give clear guidance about which areas of service should be addressed, and with what resources. For instance, satisfaction with customer communications appears to be slightly lower than satisfaction with billing, field service, or call-in/walk-in service. But is communication as strong a driver of overall satisfaction as the other attributes—or is it possibly stronger? We cannot know.

There is another problem with this an overall satisfaction rating. Even though we sharpened respondents' thoughts about the categories of service by identifying specific attributes of service that are relevant to them, we cannot assume they are able to isolate only the relevant factors in their thinking. To some unknown degree, other attitudes inevitably intrude upon their judgment. At a minimum, we would expect overall attitudes toward the utility to intrude on judgments about the individual categories. (The normal assumption, of course, goes the other way around, i.e., that the overall judgment is the sum of the category judgments.) For this reason, we do not place great importance on the ratings of satisfaction with broad categories of service. Even the overall rating of satisfaction is questionable if taken by itself.

To produce a more accurate and stable measure of overall satisfaction, we performed a statistical procedure known as *common factor analysis* on the seven general satisfaction items. Common factor analysis looks for underlying patterns among correlations between the variables.[9] In this case, the general satisfaction items were all moderately highly correlated with one another, meaning that registering high overall satisfaction, for instance, tends to accompany high satisfaction with billing, price, field service, and the rest of the general categories of service. The factor analysis procedure produced a single, strong factor with a combined correlation among all the variables of 0.88. This implies there is indeed a single underlying dimension of satisfaction and dissatisfaction with the utility, and confirms our suspicion that a general attitude formed about the utility as a whole colors all of the ratings.

Although responses to the overall satisfaction rating make it appear that all utilities are achieving approximately the same level of satisfaction, the common factor analysis reveals greater differences. The factor analysis procedure creates a new variable, representing what might be termed "underlying" customer satisfaction with the utility. This variable is "normalized," so that the mean value among all respondents from all samples is equal to zero, and the standard deviation is 1 (about 68% of the respondents will have underlying satisfaction scores between −1 and +1). Combining several satisfactions ratings into a single measure uncovers more variation between utilities than with the single item measure. Table 3.1 shows a comparison of overall satisfaction ratings for the five samples, using the single survey item and using the "underlying" satisfaction factor score. The proper comparison is between the normalized survey item and the factor score, because these have the same scale.[10]

---

[9] Correlation is a measure of association between two variables. For this analysis, we used a statistic called Pearson's r, the value of which varies between −1 and +1. A correlation of either +1 or −1 between two satisfaction items would mean that knowledge of a person's satisfaction on one item would provide perfect knowledge of their satisfaction on the other. A correlation of 0 would mean knowledge of satisfaction on one item gives us no information about satisfaction on the other. For correlations that are not near 0 we can say that knowledge about one item tends to tell us something of the other, with the strength of the relationship greater as the correlation approaches −1 or +1.

Technical note: We used maximum likelihood estimation to extract a single common factor. The correlation between the individual items and the factor are as follows: Overall Satisfaction 0.76, Price 0.61, Water Quality 0.65, Billing 0.75, Field Service 0.76, Call-in/Walk-in Service 0.73, Customer Communication 0.70.

[10] "Normalizing" the measure entails subtracting the mean from each score (in this case each utility average) and dividing by the standard deviation. The single survey item has a standard deviation of 21.08. The factor score is already normalized when it is calculated. By normalizing the single survey item score, the two values become directly comparable in terms of standard deviations from the mean.

Table 3.1
Comparison of overall satisfaction among five samples

Sample	Average Rating: Single Survey Item	Average Rating: Single Survey Item (Normalized)	Average Rating: Factor Score (Normalized)
Kissimmee	82.40	0.11	0.26
National	81.12	0.05	0.10
Nashville	81.10	0.05	0.04
Grand Average	**80.12**	**0.00**	**0.00**
Phoenix	79.70	-0.02	-0.08
Washington D.C.	76.03	-0.19	-0.38

Using either method, the order of the samples from highest to lowest overall rating is identical. The factor score method, however, spreads the ratings out more. For instance, Kissimmee's rating is farther from the mean (0.26 vs. 0.11), and Nashville is no longer essentially tied with the national sample but falls below it. This indicates greater underlying differences among the samples than is evident using the single item. The reason for the difference is that the factor score approach includes more information about customers' attitudes. Thus it provides a firmer basis for later estimates of the importance of different attributes.

Satisfaction with Current Performance on the Attributes of Service

A more useful set of findings for decision-making purposes is customers' current satisfaction with the 17 attributes of service. These have the advantage of referring to specific, concrete experiences respondents have had. Moreover, they refer to aspects of service that managers can improve on or economize, whereas the general satisfaction items do not give clear direction about what changes need to be made. Because they are more concrete, we believe the attribute ratings are more independent of overall judgments about the utility. Even so, they are not immune from the problem that overall judgments color judgments about the specific attributes.[11]

It is important to keep in mind that for this research we measured current satisfaction only among people who had direct experience with a given attribute of service. So, for instance, only people who had a bill problem were tallied for the attribute Resolution of Bill Problems; only people who had a field representative come to their home were tallied for the attributes Appointment Window, Professionalism, etc.

Another important point is that satisfaction was measured indirectly. We did not ask *how* satisfied customers were with a given attribute of service. Rather, we first asked them to tell us how satisfied they *would be* with different levels of service, and then asked them which level of service was closest to the one they actually receive. We took the prior rating of that level as their

[11] Common factor analysis confirms this assertion of both a basic independence of the attributes and their relationship to underlying satisfaction. Analysis of all general satisfaction items together with all attributes shows that the general items form a distinct factor of their own and the attributes form three separate factors, but that the factors are themselves correlated with one another.

current satisfaction. This approach has both disadvantages and advantages. If respondents cannot find a level of service that perfectly matches what they think they have received, we can assume that the satisfaction rating of the nearest level is somewhat off target. The advantage of the approach, however, is that it first brings different hypothetical levels of service to mind, and then it derives the satisfaction rating from a very concrete recollection of experiences. This is not to say that satisfaction with the experience does not also color recollection of the level of service received.

Mean satisfaction with each attribute of service, for each sample, is presented in Table 3.2.

Table 3.2
Mean current satisfaction with seventeen attributes of customer service

	Sample					
	Kissimmee	Nashville	Phoenix	Washington D.C.	National	Average
Bill problem resolution	79	81	82	68	81	78
Billing schedule	91	91	95	68	65	82
Hours of operation: field service	72	71	71	68	68	70
Appointment window: field service	74	70	67	63	64	68
Response time: field service	71	71	70	61	67	68
Arrival etiquette: field service	83	80	84	75	81	81
Professionalism: field service	83	81	82	73	81	80
Follow-up: field service	80	75	73	66	74	74
Hours of operation: customer service	68	71	69	66	77	70
Authority of reps: customer service	86	80	84	76	82	82
Professionalism: customer service	86	82	82	69	83	80
Automated phone: customer service	72	62	64	55	65	64
Phone hold time: customer service	79	72	76	61	81	74
Line wait: customer service	80	76	73	64	81	75
Frequency utility communicates	74	72	71	67	68	70
Water quality info schedule	68	68	68	66	64	67
Internet information	69	61	66	62	56	63

A figure providing clearer visual comparison of average satisfaction rates (across all samples) among the seventeen attributes is presented in Figure 3.2.

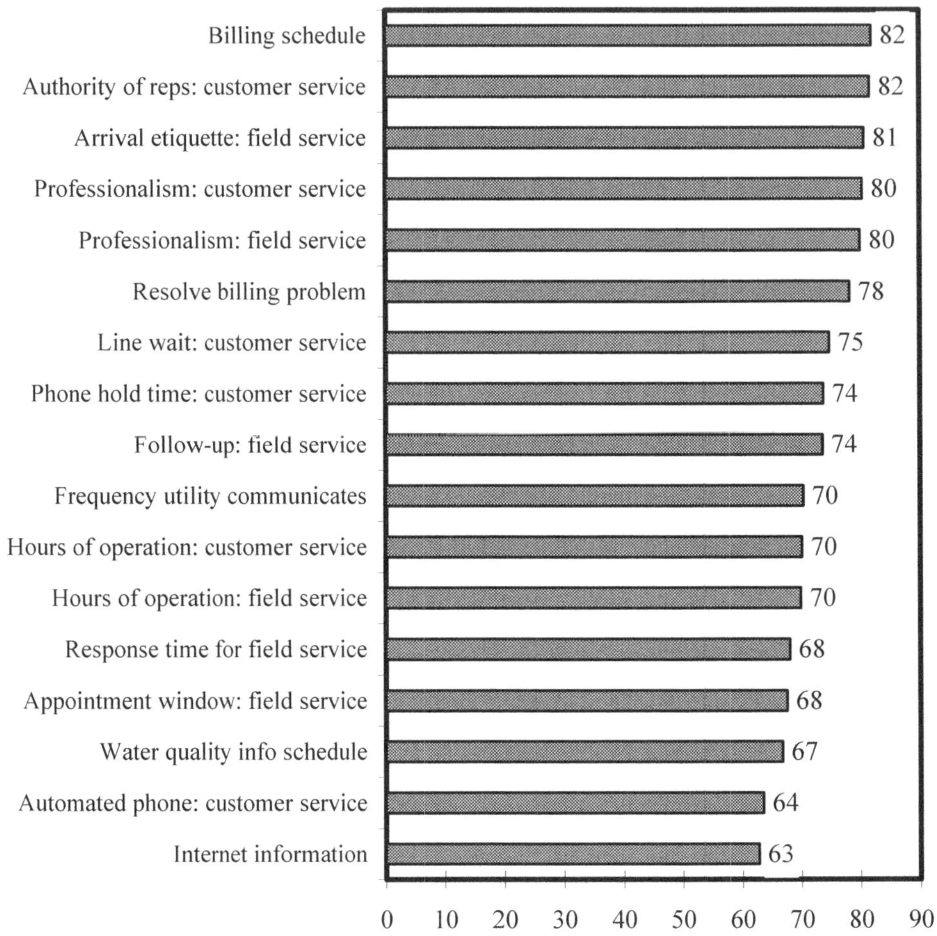

Figure 3.2 Comparison of average satisfaction among seventeen attributes of service

The highest average level of satisfaction (82) was recorded for the attributes Billing Schedule and Authority of Representatives, which asks about the representatives' ability to make decisions. Most customers receive their bill monthly and rated this as the highest level of service. D.C. WASA customers at the time of the survey received their bill quarterly, and rated that service highest, but rated it lower than the other samples rated monthly billing frequency. The next highest level of satisfaction (81) was recorded for Arrival Etiquette of field service representatives, primarily asking about their timeliness.

Consistent with what we heard in the focus groups, the lowest satisfaction (63) was recorded for the attribute related to Internet Information. Few people use the Internet to contact their utilities, and most people said their utility offers a relatively low level of service. The next lowest satisfaction was recorded for an attribute related to automated phone systems, known from the focus groups to be a sore subject among customers. Also getting low ratings were Water Quality Information Schedule (67), Appointment Window for field service (68), and Response Time for field service (68).

This kind of information, which would be typical of high-quality customer satisfaction research, might lead managers to the conclusion that the attributes of service with the highest satisfaction ratings are currently being provided at adequate levels, and that attention should be given to attributes with the lowest ratings. That could be a mistaken judgment, however. For instance, if any of the highly rated attributes are particularly important in the formation of overall judgments of satisfaction, and if they are relatively inexpensive attributes for which to improve service, they might be the very ones to address. Conversely, low-rated attributes may not be important in the formation of overall satisfaction, and they may be expensive to address. It is possible that some attributes of service are being offered at a level that is too high, and that overall satisfaction could be improved if resources could be diverted to other attributes that have a greater impact on overall satisfaction.

Another problem is that information about current satisfaction gives utility decision makers no way of knowing at what level satisfaction ratings should begin to cause concern. This question becomes especially confusing for attributes with middling satisfaction ratings. Is a rating of 75 for Line Wait for in-person customer service good, or bad? Trends across time or comparisons with other utilities, if available, would give some guidance; but even Kissimmee, which scored significantly above average, might benefit from improvement. The optimal decision to address an area of customer service depends upon three factors:

- How quickly satisfaction rises or falls as service improves or declines
- The importance of the attribute to overall satisfaction, and
- The cost of making improvements.

SATISFACTION AT DIFFERENT LEVELS OF SERVICE

A central part of the method employed was asking survey respondents to rate how satisfied they would be at different levels of service—high, medium, and low—for the *same* attribute.[12] The service levels were designed to capture realistic possibilities. We made an effort to include levels that were close to levels customers actually receive, with the constraint that we wanted to develop a single instrument for all partner utilities. So, for instance, we did not specify times of day for customer service but only total hours of operation with a parenthetical example of specific times. Where possible, we included levels of service above and below what customers currently receive, so there would be room for improvement and economizing in the modeling of optimal service.

Graphing the average satisfaction at each level of service produces a preference contour. This represents the rate that satisfaction drops as service levels decline or rises as service levels increase. It must be remembered that the levels of service are at best ordinal, so the slope of the curves cannot be compared from one attribute to another in any straightforward way. A weighting procedure to make such comparisons possible will be discussed below.

In this portion of the report, we refer to "projected satisfaction" because survey respondents were estimating the hypothetical satisfaction they would feel if they received different levels of service. There is evidence that customers can accurately estimate such values

[12] The levels were presented in what was presumed to be a decreasing order of service. In most cases, this order of service levels is quite obvious and implicit, for instance decreasing wait times represent increasing levels of service. In only three cases was the presumed order disconfirmed by survey responses. Fortunately, the modeling of optimal service levels does not require that the assumed order is the correct one. Moreover, there was some concern that the order in which service levels were presented would bias responses about satisfaction. The fact that there were three cases where survey respondents contradicted our assumption suggests that the question design did not influence respondents unduly.

(Oliver 1980). The results of our research, moreover, indicate that responses to the different attributes differed from one another in ways suggesting an underlying attitude rather than a kind of unconscious reaction to the survey task itself—what psychologists call a "response set." The shapes of the curves differ substantially from one service attribute to another, and the few attributes where one sample does not follow the general pattern also suggest an intelligent matching of responses with real circumstances. Nevertheless, the possibility of systematic bias in the rating of satisfaction, for instance under- or over-estimating satisfaction associated with a given wait time, cannot be ruled out and would require supplemental research to uncover.

A Remarkable Consistency of Preferences Across Samples

To our knowledge, no one has ever conducted a survey of this design simultaneously in multiple service territories, using an identical survey instrument. The research team was anticipating that there might be interesting differences in the preference contours of different locations, reflecting the unique culture or experience of the locations, and we were eager to see if we could make sense of these differences. What we did not anticipate, however, is what we actually found: that the preference contours of all partner utilities and the national sample were nearly identical for 14 of the 17 attributes, and for the remaining three attributes either a single utility or the national sample accounted for most of the divergence.[13]

Billing

The service category *Billing* had two attributes, Resolution of Bill Problems and Billing Frequency. In the original design, an attribute called "clarity of bill information" had been included. This attempted to measure whether customers could read and understand the information presented on their bill. The focus group research did not reveal any significant problems in this regard among the partner utilities, however, and it is difficult to calculate costs for levels of service. We therefore dropped the attribute from the study. If utilities had reason to expect their bills are not easy to understand, they may want to include such an attribute in their model.

---

[13] To put some numerical precision on this assertion, we did an analysis of variance for each level of each attribute, testing the hypothesis that the samples were different. For fourteen of the attributes, the hypothesis that the samples are different was rejected in 90% of comparisons (466 out of 520). In other words, 90% of the time the samples' satisfaction scores on the attributes were too similar to ascribe differences to the underlying population. (The samples each had between 440 and 678 observations, so a difference would have to be about 5 rating points for it to be statistically significant.) For the attribute Billing Frequency, 27 out of 40 comparisons showed statistically significant differences, but sixteen of those comparisons were with a single utility, D.C. WASA. For the Customer Communication attributes dealing with the frequency and scheduling of information, 28 out of 80 comparisons were statistically significant. The majority of these statistically significant comparisons were with the national sample. These divergent cases are discussed in the text.

Resolution of Bill Problems

Our focus group research indicated that most customers do not expect to have any problems with their water bill. Among those few who had experienced problems, however, some reported satisfaction with the way the problem had been handled, while others were quite frustrated and apparently angry about the experience.

For the survey, we designed an item with three levels of service for the hypothetical situation where a customer has a problem with his or her bill.

Level 1	The billing problem is resolved after one contact with the utility and with no negotiations
Level 2	The billing problem is resolved after two or three contacts with the utility and minor negotiations
Level 3	The billing problem is resolved after more than 3 contacts with the utility and difficult negotiations

The responses from each utility are represented in Figure 3.3.

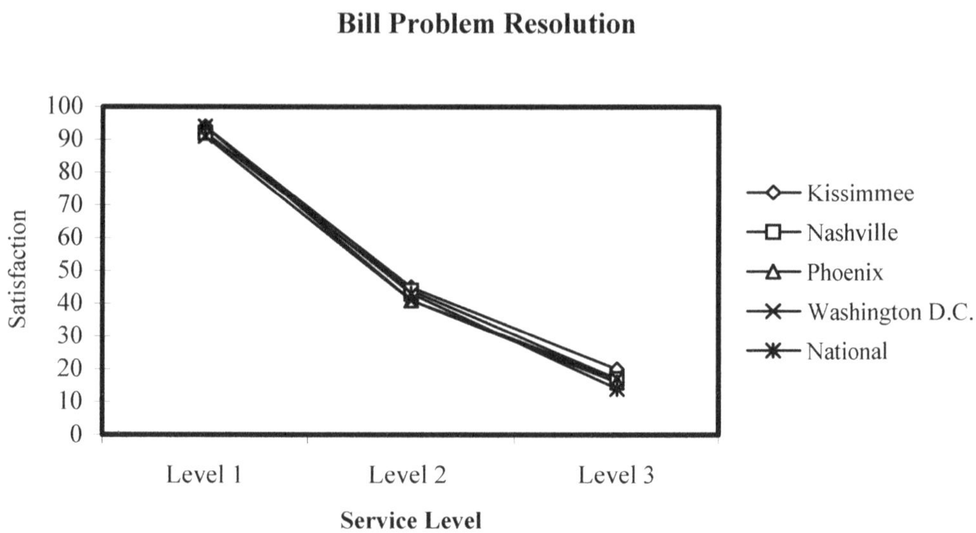

Figure 3.3 Projected satisfaction with bill problem resolution in five customer samples

Among the partner utilities and the national sample, projected satisfaction with this attribute averaged 93 points for Level 1 service, dropped steeply to 43 for Level 2 service, and dropped again, though somewhat less steeply, to 17 for Level 3 service. Clearly, customers much prefer that a problem be solved with only one contact. It is interesting, however, that the Level 3 rating, though clearly representing a very low level of satisfaction, is by no means the lowest level that was registered. For instance, the experience of rudeness from a utility representativeness yielded satisfaction scores of approximately 5.

Comparing people who reported they experienced a bill problem with the entire sample, we find that this group rates satisfaction somewhat lower for Level 1 service (88), but rates Level 2 and Level 3 service (47 and 22, respectively) higher than the entire sample. One might speculate that people who have *not* had a problem with their bill anticipate that negotiations with the company will be more troublesome than in fact they are. Or, if a few customers regularly have bill problems they may become more comfortable in their negotiations with the utility.

Billing Frequency

The focus groups did not reveal any significant concerns about the frequency of customers' billing. At the time of this research, all partner utilities except D.C. WASA were billing monthly. D.C. WASA was at that time billing quarterly, although they are now in transition to monthly billing. Bill frequency was included as an attribute because our past experience suggested this might be an aspect of service where savings could be achieved, by reducing service to less frequent billing. We wanted to be able to test that possibility.

The item measuring this attribute had four levels of service.

Level 1	The water bill is sent every month
Level 2	The water bill is sent every two months
Level 3	The water bill is sent every three months
Level 4	The water bill is sent every six months

For this attribute, it is unclear that customers will consider less frequent billing a lower level of service. That possibility does not affect the model, however. The responses from each utility and the national sample are represented in Figure 3.4.

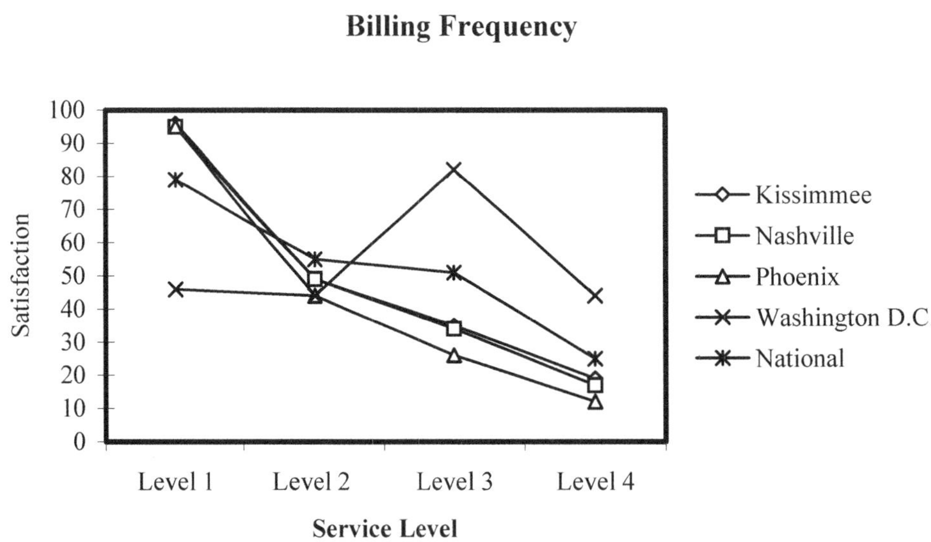

Figure 3.4 Projected satisfaction with billing frequency in five customer samples

Of all the attributes, this one has the least similarity among the different satisfaction profiles. The D.C. WASA sample is the most divergent. It is important to note again that this was the only utility that was not billing customers monthly at the time of the survey. Rather, they billed customers every three months, the level that was given the highest satisfaction rating. Other service levels are all given equivalent, middling ratings, representing the fact that people had diverse, offsetting preferences for the other billing schedules. Standard deviations for these ratings were unusually high, indicating less than normal agreement about which alternative level of service is best. Customers from the other partner utilities estimated similar preferences, with Phoenix preferences dropping off slightly more quickly between Level 1 and Level 2 service. The national sample reflects a similar pattern, perhaps because it includes more diversity of actual experience—people billed at rates other than once per month, which flattens out the curve.

The aberrant D.C. WASA finding suggests a tentative conclusion, which is well supported by the social-psychological literature: people tend to prefer what they are familiar with. They may resist change where the change is not generally and self-evidently perceived as an improvement in service. For most attributes in this research, this was not an issue. The order of service levels was clearly agreed upon by all samples, no matter what level of service they were currently receiving. With this attribute, however, we can be less sure that projected satisfaction reflects what would actually be experienced, once customers become familiar with a new level. It is noteworthy that the D.C. WASA sample, though giving quarterly billing their highest rating, did rate it significantly lower (82) than the other partner utilities rated monthly billing (average 95).

Field Service

The service category Field Service had six attributes of service: Hours of Operation, Appointment Window, Response Time, Arrival Etiquette, Professionalism, and Follow-Up. As in the case of Billing, the focus group research eliminated several attributes we had considered using. For instance, in prior versions of this service, we had included items for the restoration of property after service, and for the clear labeling of trucks and uniforms with the utility logo. Neither of these attributes was given any importance in the focus groups discussions. Nevertheless, we would expect that a utility performing well below expectations on these attributes of service—especially the restoration of property—could expect to see an impact on customer satisfaction.

Hours of Operation

The focus group research did not reveal a great deal of concern about the hours during which non-emergency field service is available. A number of respondents spoke well about the fact that utility personnel had responded quickly when they had an emergency. Regular and emergency services are not clearly distinguished in the minds of customers, apparently. All of the partner utilities currently offer 24-hour emergency service, and it is not uncommon that utilities use a rather liberal definition of what constitutes an emergency, at least where an initial investigation is concerned. Thus, customer perception of the hours of operation for field service will be determined by the kind of call they make.

The item measuring this attribute had four levels of service.

Level 1 Field service (non-emergency) is available 24 hours/day, 7 days/week
Level 2 Field service (non-emergency) is available 12 hours/day (for example, 8 a.m. to 8 p.m.), 7 days/week
Level 3 Field service (non-emergency) is available 8 hours/day (for example, 9 a.m. to 5 p.m.), Monday–Friday
Level 4 Field service (non-emergency) is available 4 hours/day (for example, 7 a.m. to 11 a.m.), Monday–Friday

The responses from each utility and the national sample are represented in Figure 3.5.

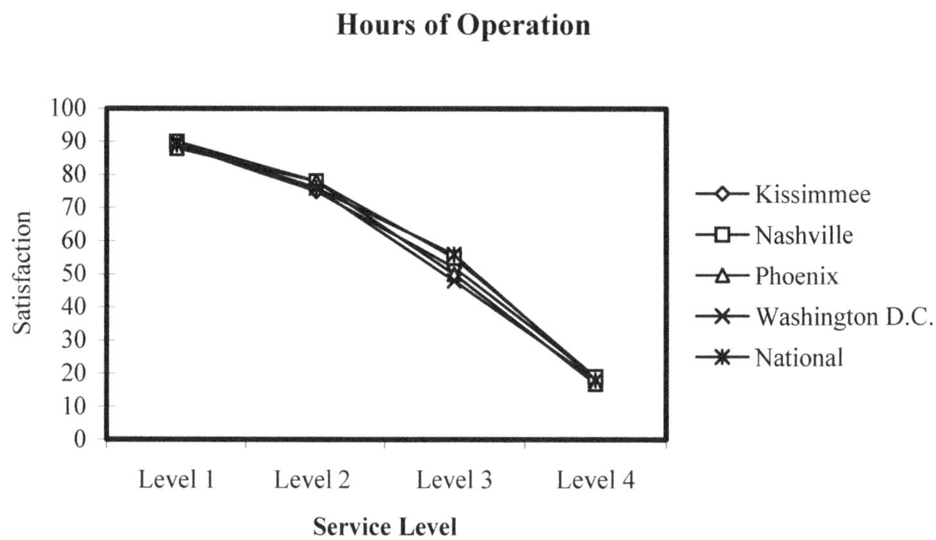

Figure 3.5 Projected satisfaction with field service hours of operation in five customer samples

It is interesting that ratings of the highest level of service are at or below 90 (average 89). This is surprising in light of the fact it is hard to conceive service better than 24 hours per day and seven days per week. Other attributes have Level 1 ratings in the order of 98 points, so this suppression of satisfaction does not reflect a feature of the rating task itself. Rather, it appears to reflect the fact that the question begins with a premise that the customer has a problem that requires service. Thus, the question already assumes less than optimal service—or at least an unpleasant problem—from the customer's perspective. There is a similar suppression of Level 1 ratings on the attribute "Resolution of Bill Problems" and others where the assumption is that one has already encountered a problem.

The decline in service rating between levels 1 and 2 is relatively shallow, on average 12 points (to 77), suggesting that 12 hours of field service per day is acceptable to many customers. Where service dips below the hours of the workweek, however, and customers anticipate the inconvenience of taking off work to deal with service representatives, satisfaction begins to decline considerably (to an average of 52). But it is only when service hours drop to 4 hour per day on weekdays that customers really assert their displeasure, rating service an average of 18 points.

Appointment Window

We did hear a number of complaints from focus group participants about the period of time people are required be at home, waiting for field service representatives to arrive. These were often accompanied by comments that the service representatives often arrived at the very end of the appointment window, or were even late. The comments primarily came during the generic discussion of customer service, prior to the discussion of water utilities per se. Nevertheless, this did appear to be an important attribute of service for the model.

The item measuring this attribute had three levels of service.

Level 1	The appointment is scheduled within a 2-hour window
Level 2	The appointment is scheduled within a 4-hour window
Level 3	The day of the appointment is scheduled, but not a time during the day

The responses from each utility and the national sample are represented in Figure 3.6.

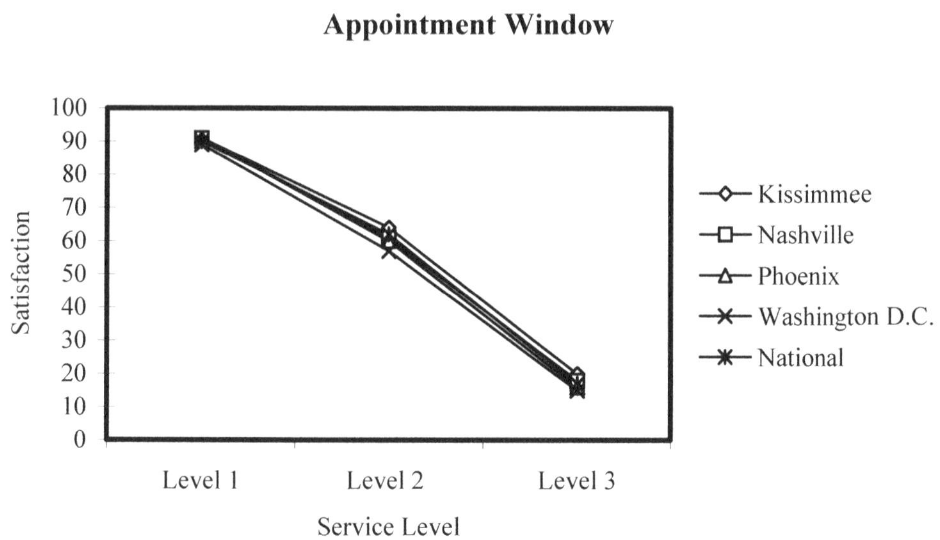

Figure 3.6 Projected satisfaction with field service appointment window in five customer samples

In the focus groups there was some disagreement about what length of time was acceptable for the appointment window, some saying two hours and others saying fours hours. It is clear that, given a choice, a two-hour window is strongly preferred over a four-hour window. It is not surprising, however, that satisfaction falls even more steeply when the window increases to a full day.

Response Time

The focus group research revealed several complaints about the response time of field representatives, but these were for businesses other than water utilities, most notably television cable companies and in a few cases electrical service. Response time did not emerge in the discussion of water utilities; however, it clearly can become a problem if wait times are excessive. This is especially true if response-time problems are compounded by other service problems. For instance, cable companies typically were faulted for service reliability problems in the same breath as slow service response times: "My service keeps going out and it takes the company forever to fix it."

The item for this attribute had four levels of service.

Level 1	The service representative arrives within 4 hours after your call for service
Level 2	The service representative arrives on the same day as your call for service
Level 3	The service representative arrives the next day after you call for service
Level 4	The service representative arrives within 3 days after your call for service

The responses from each utility and the national sample are represented in Figure 3.7.

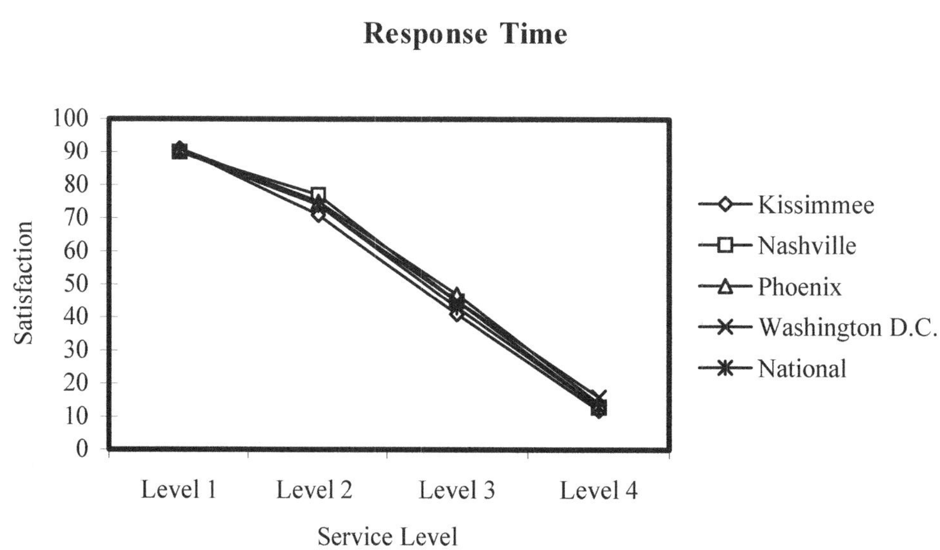

Figure 3.7 Projected satisfaction with field service response time in five customer samples

Clearly, customers strongly prefer to receive service on the same day that they call. This may not be a realistic option for many utilities. In the current research format, we could not test the effect on satisfaction of every possible service response. For instance, we are aware some utilities provide an initial investigative response on the day a call for service is made, with a subsequent visit for repair on a later date, if needed. Future research might disentangle the satisfaction impact of such a policy.

Arrival Etiquette

As mentioned earlier, we heard a number of comments in the focus groups about field representatives arriving late. This appears to be a more important service consideration that the actual size of the appointment window. Because the complexity of service scheduling makes a certain number of late calls inevitable, we wanted to test the effect on satisfaction of phoning customers to inform them when a customer service representative will be late. It does not appear that this is a typical practice, however. To our knowledge, most utilities do not even keep track of the percentage of time that service representatives require customers to wait beyond the appointment window. This would be useful information in responding to findings on this attribute of service.

The item for this attribute had four levels of service.

Level 1	The service representative arrives on time and phones beforehand to let you know they are coming
Level 2	The service representative arrives on time without phoning beforehand
Level 3	The service representative arrives late but phones beforehand to let you know
Level 4	The service representative arrives late without phoning to let you know

The responses from each utility and the national sample are represented in Figure 3.8.

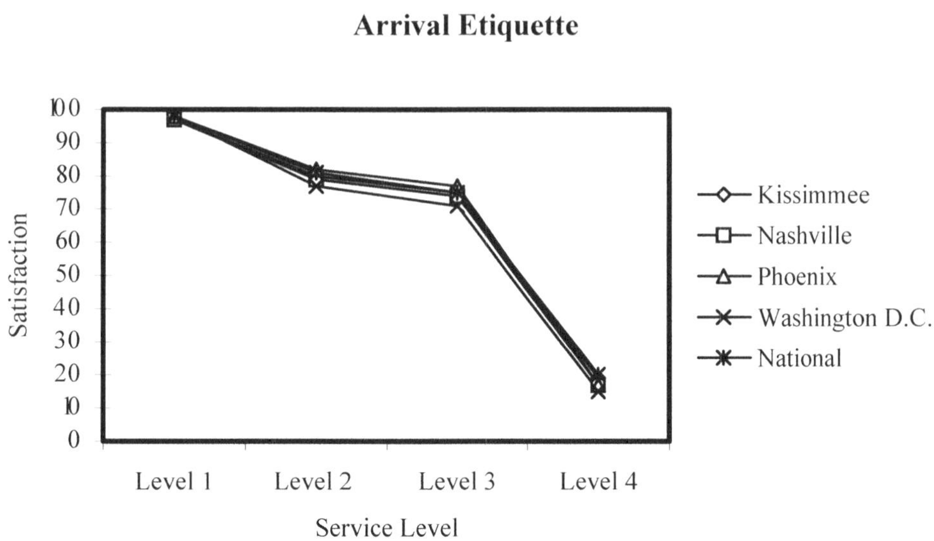

Figure 3.8. Projected satisfaction with field service arrival etiquette in five customer samples

Clearly, phoning customers to keep them apprised of arrival time is a greatly preferred practice. It is especially important to notice that satisfaction dips very little between Level 2 (average 80) and Level 3 (average 74). In other words, customers are almost as satisfied when representatives are late, *as long as they phone ahead to let customers know they will be late*, as they are with the "normal" practice of arriving on time without a prior phone call. Not surprisingly, arriving late without a call is a greatly disfavored level of service.

Professionalism

An experience that brought some of the strongest statements of dissatisfaction in the focus groups was a lack of professionalism on the part of business representatives that customers deal with. The most common complaint was that representatives do not listen to what customers are saying. Another complaint was that representatives do not explain what they are doing or what organizational procedures they are following. The strongest complaints were voiced for situations where a representative is rude or impolite to customers.

We designed a single item to capture all three of these aspects of professionalism. In most instances, it would be preferable to design a separate item for each aspect. In this case, however, it was too difficult to design multiple levels of service for each aspect of professionalism that would be meaningful in the context of a pencil-and-paper survey. It would also have been difficult to assign different costs to these levels. Moreover, we were under significant constraints to keep the total length of the survey manageable. Since these three aspects of service were all discussed together in the focus groups, as if they were part of a common phenomenon, we combined them into a single attribute.

The item for field service professionalism had four levels of service.

Level 1	The service representatives give clear information about what they are doing, listen carefully to what you say, and are polite and friendly
Level 2	The service representatives do not give clear information about what they are doing, but listen carefully to what you say, and are polite and friendly
Level 3	The service representatives do not give clear information about what they are doing, do not listen carefully to what you say, but are polite and friendly
Level 4	The service representatives do not give clear information about what they are doing, do not listen carefully to what you say, and are a little rude or unfriendly

The responses from each utility and the national sample are represented in Figure 3.9.

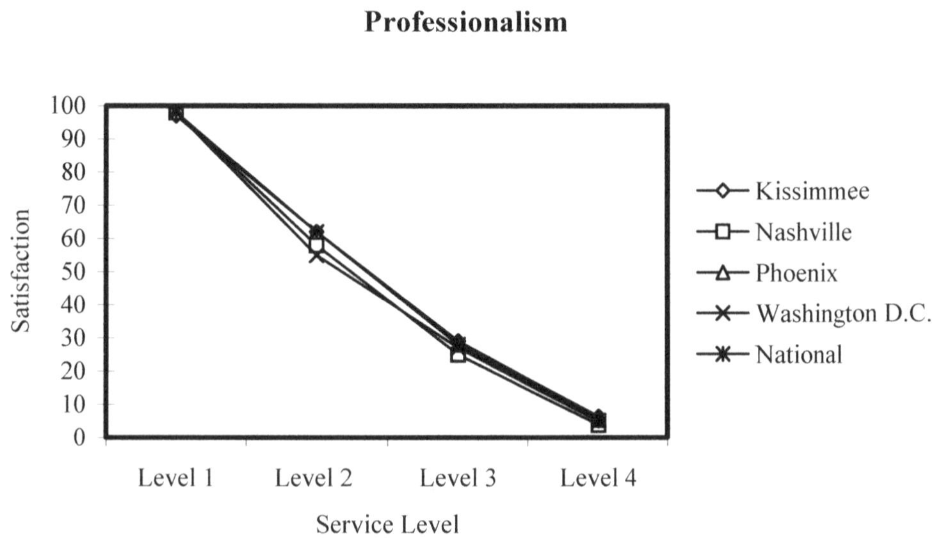

Figure 3.9 Projected satisfaction with field service professionalism in five customer samples

This attribute had the greatest range between Level 1 and Level 4 service, with average ratings of 98 and 5, respectively. The first two decreases in service each brings a drop in satisfaction of 35 points, and one suspects that respondents may have essentially run out of points to subtract for the final reduction in service. At any rate, this appears to be a very important attribute of service, reinforcing the focus group evidence that people are quite sensitive to the way they are treated by utility personnel.

Follow-up

We are not aware of any water utility that regularly follows up on each service call to assess whether the customer was satisfied. This is an increasingly common practice in other industries, however, and we wanted to test whether it would have any impact on customer satisfaction. We also wanted to see if following up in writing would bring as much satisfaction as following up by phone. The focus groups found mild support for the idea of follow-up calls. For the most part, it appeared that the focus group participants did not expect any problems with the service they receive, so did not anticipate the need for follow-up calls.

The item for Field Service Follow-up had four levels of service.

Level 1	The service representative follows up in writing and with a phone call
Level 2	The service representative follows up with a phone call but not in writing
Level 3	The service representative follows up in writing but not with a phone call
Level 4	The service representative does not follow up

The responses from each utility and the national sample are represented in Figure 3.10.

Follow-up

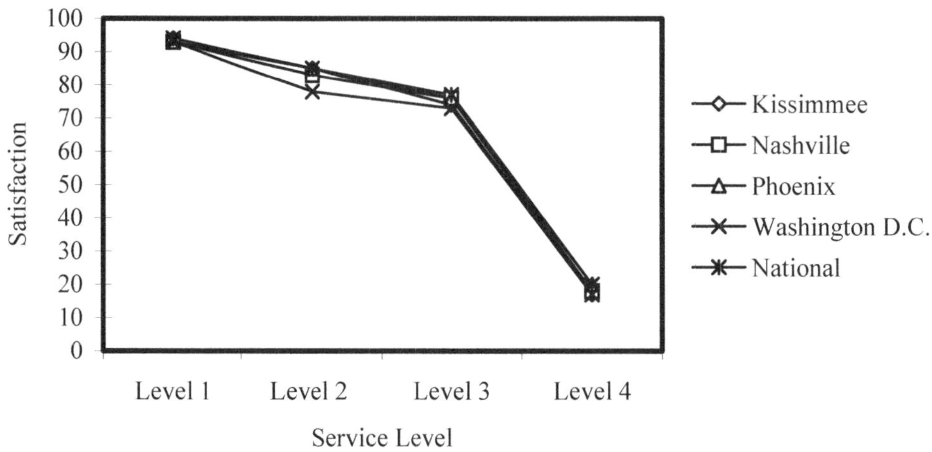

Figure 3.10 Projected satisfaction with field service follow-up in five customer samples

Contradicting what the focus groups led us to believe, the survey respondents reported high levels of satisfaction for field service follow-up, with a slight preference for follow-up by phone (average 83) compared to follow-up by mail (average 75). To our surprise, a fairly high percentage of respondents reported they had actually received a follow-up call for service at their property, despite the fact that the utilities say they do not routinely make such calls. There are, of course, errors in recall associated with such answers, and a strong degree of good will toward a utility could influence respondents to recall follow-up efforts that had not actually taken place. In other words, following up with customers may be a good way to build or solidify good will.

Call-in and Walk-in Service

The service category Call-in and Walk-in Service had six attributes: Hours of Operation, the Authority of Representatives (i.e., to make decisions), Professionalism, Automated Phone System, Phone Hold Time, and In-person Wait Time.[14] Although this clearly is an important category for customer service managers, the focus groups spent relatively little time talking specifically about service provided by water utilities. They, of course, did have many comments to make about customer service in general, many of which can be translated directly into items for the survey.

Hours of Operation

The focus group research did not uncover any significant issues surrounding the hours of operation of customer service. We did not anticipate this would be an area where improvement is required; however, there are significant costs associated with the hours of service, and we

[14] We use this admittedly awkward label "Call-in and Walk-in Service" for the category, rather than the more succinct "customer service," because we want the latter term to be more inclusive, taking into account every way a water utility comes into contact with its customers.

thought it might be an area where savings would be possible. Our chief expectation was that satisfaction might decline considerably if customer service was offered only during regular business hours.

The item for this attribute had four levels of service.

Level 1	Customer service is available 24 hours/day, 7 days/week
Level 2	Customer service is available 12 hours/day (for example, 7 a.m. to 7 p.m.), Monday–Friday, and some weekend hours
Level 3	Customer service is available 12 hours/day (for example, 7 a.m. to 7 p.m.), Monday–Friday
Level 4	Customer service is available 8 hours/day (for example, 9 a.m. to 5 p.m.), Monday–Friday

The responses from each utility and the national sample for the attribute Customer Service: Hours of Operation are represented in Figure 3.11.

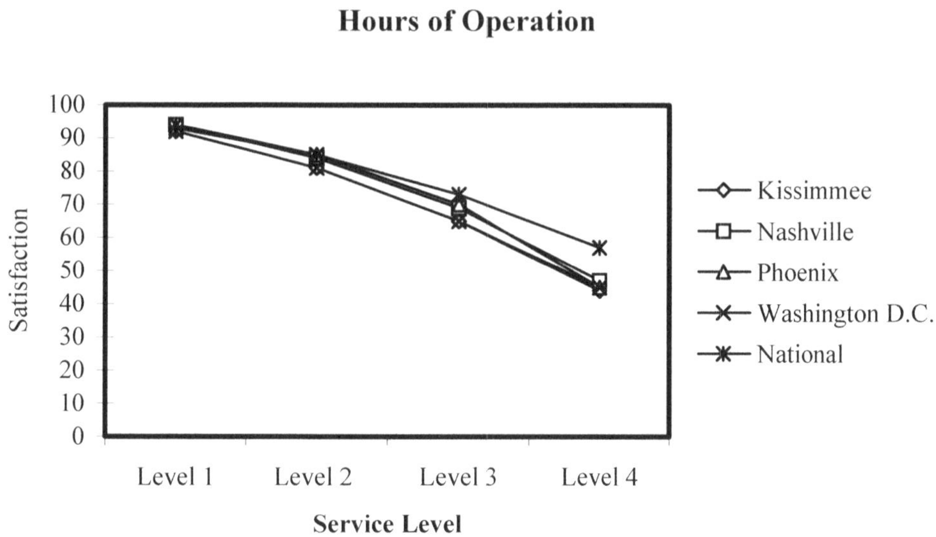

Figure 3.11 Projected satisfaction with customer service hours of operation in five customer samples

Consistent with our expectations, this attribute had one of the shallowest preference contours of all the attributes we tested. Satisfaction with Level 4 service is remarkably high. It is also notable that the national sample diverges from the others, 12 points above the average of the other four samples at Level 4. We do not have an adequate explanation for this pattern. The average perceived level of current service for the national sample is lower than for the partner utilities, although by only a small amount (2.89 versus 2.54, respectively—remember that the larger the number the poorer the service). This may, once again, reflect a degree of habituation to the level of service one is used to. We would expect greater dissatisfaction with lower-than-current service than with current service.

Authority of Representatives

In the focus groups, there was general familiarity with the circumstance where customer service representatives do not have sufficient authority to make decisions, thus complicating and slowing transactions and increasing the perception of inefficiency in the organization. We combined this observation with a more general concern that typically accompanied it about the general competency of the representative. To test the importance of this attribute, we designed an item with three levels of service.

Level 1	The customer service representative is authorized to make decisions and completes the job easily
Level 2	The customer service representative has to seek authorization from others but completes the job easily
Level 3	The customer service representative has to seek authorization from others and completes the job with difficulty

The responses from each utility and the national sample are represented in Figure 3.12.

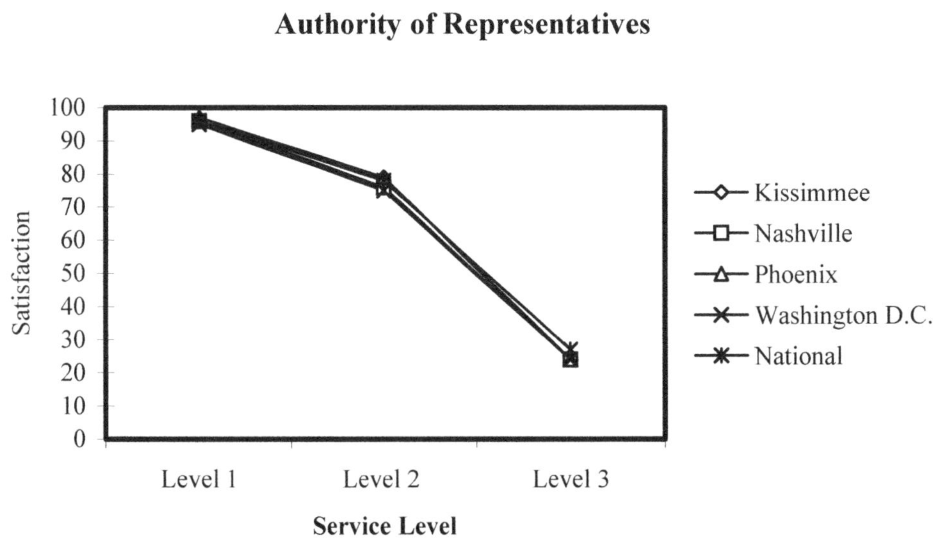

Figure 3.12 Projected satisfaction with customer service authority of representatives in five customer samples

The findings suggest that seeking authorization from someone else is not a problem in and of itself, but much more of a problem when it creates the impression of incompetence, which is the overriding concern.

Professionalism

The logic of this attribute is no different than it was for the category field service. In the minds of customers, there is no clear distinction between the two sets of utility representatives. The reason for having two items is that we expected the costs of increasing or decreasing service to be different. Again, the item compounds three aspects of professionalism that appear to be closely related.

The item for professionalism had four levels of service.

Level 1	The service representatives give clear information about what they are doing, listen carefully to what you say, and are polite and friendly
Level 2	The service representatives do not give clear information about what they are doing, but listen carefully to what you say, and are polite and friendly
Level 3	The service representatives do not give clear information about what they are doing, do not listen carefully to what you say, but are polite and friendly
Level 4	The service representatives do not give clear information about what they are doing, do not listen carefully to what you say, and are a little rude or unfriendly

The responses from each utility and the national sample are represented in Figure 3.13.

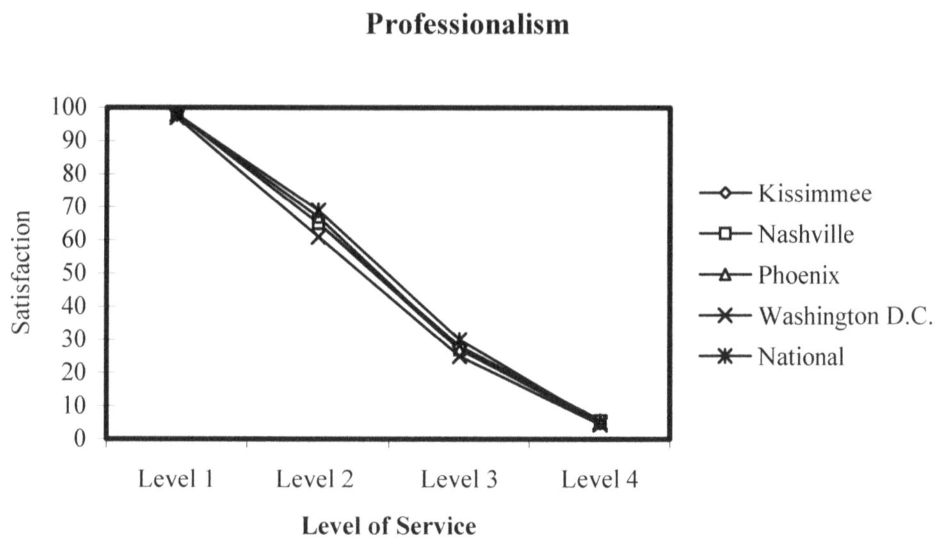

Figure 3.13 Projected satisfaction with customer service professionalism in five customer samples

The most interesting finding with regard to this attribute is that the average drop between Level 2 and Level 3 is larger than for the equivalent item for field service (a 38-point drop-off and a 33-point drop-off, respectively). Although the difference is small, it is statistically significant and possibly reflects the fact that, in the context of walk-in and phone-in customer service, customers perceive themselves to be at less of a disadvantage with respect to technical expertise. It reinforces a point that came though very powerfully in the focus groups, that customers strongly want to feel they are being listened to.

Automated Phone System

Complaints about automated phone systems were both frequent and heated in the focus groups. This clearly is a "sore" point with many consumers. Once again, however, these complaints were mainly voiced in the initial, general discussion of customer service. We did not record any significant complaints about water utilities *per se*. The two main issues surrounding such systems was the number of inputs and the number of options from which customers have to select at each input point.

Clearly, this attribute could be defined on numerous different levels. We chose the following three.

Level 1	An automated system answers the phone. You are able to reach a customer service representative with one input from a short list of options
Level 2	An automated system answers the phone. You are able to reach a customer service representative after three inputs from a long list of options
Level 3	An automated system answers the phone. You cannot reach a customer service representative

The responses from each utility and the national sample are represented in Figure 3.14.

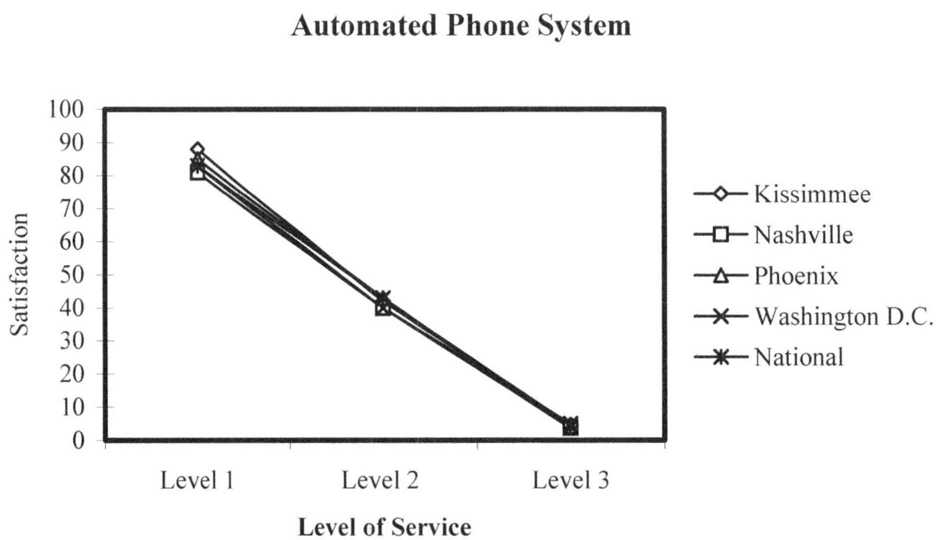

Figure 3.14 Projected satisfaction with customer service automated phone system in five customer samples

In previous research, we have defined Level 1 as reaching a live person immediately without intervention of an automated phone system. That option does seem to be preferred, and may have pulled the uppermost satisfaction rating above the mid 80's. In designing this survey, however, the partner utilities were in general agreement that the live response option was no longer viable, so we eliminated it from the survey. Level 4 service received an overall satisfaction rating of 4, lowest of any attribute in the study.

Phone Hold Time

The length of time that people must hold on the phone before they reach a customer service representative was not significant topic of conversation in the focus groups. We did not assume this was because people are insensitive to it. We thought it might be that there is rather little to say about the subject, other than the fact that waiting is unpleasant. We developed four levels of service for this attribute, each representing a different hold time. We chose the times based on input from the partner utilities, and in particular based on some benchmarking data shared by Phoenix Department of Water Services.

We are aware that the subjective perception of waiting is different than the reality. The *idea* of waiting on hold for two minutes may not seem bad—and it is the idea that customers are responding to in our survey—but the reality of a two-minute wait typically is that impatience sets in. Thus, we do not take at face value the relationship between satisfaction and time, but rather think it should be corroborated with other research. Nevertheless, we do get an indication of the rate at which satisfaction falls off as phone hold time increases.

The item for the attribute Phone Hold Time had four levels of service.

Level 1	You do not have to wait on hold. You speak to a customer service representative right away
Level 2	You are on hold for 20 seconds before speaking to a customer service representative
Level 3	You are on hold for 45 seconds before speaking to a customer service representative
Level 4	You are on hold for 120 seconds (2 minutes) before speaking to a customer service representative

The responses from each utility and the national sample are represented in Figure 3.15.

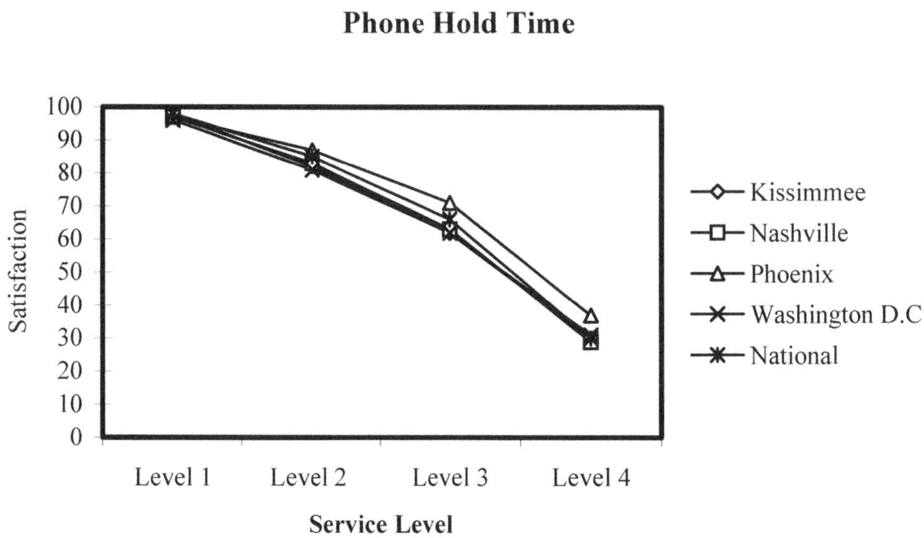

Figure 3.15 Projected satisfaction with customer service phone hold time in five customer samples

The relatively shallow profile for this attribute above Level 3 suggests that customers do not consider typical phone hold times an important source of dissatisfaction. One wonders how waiting times longer than two minutes would have been rated—although these apparently are not normally encountered by water utility customers.

In-person Wait Time

The focus groups revealed a number of complaints about waiting in line for customer service, and these complaints were directed at the local water utility, rather than being generic. Again, we based the time periods for the different levels on input from the partner utilities. The item for this attribute had four levels of service.

Level 1	You do not have to wait in line. A customer service representative serves you right away
Level 2	You have to wait in line for 1 minute before a customer service representative serves you
Level 3	You have to wait in line for 5 minutes before a customer service representative serves you
Level 4	You have to wait in line for 10 minutes before a customer service representative serves you

The responses from each utility and the national sample are represented in Figure 3.16.

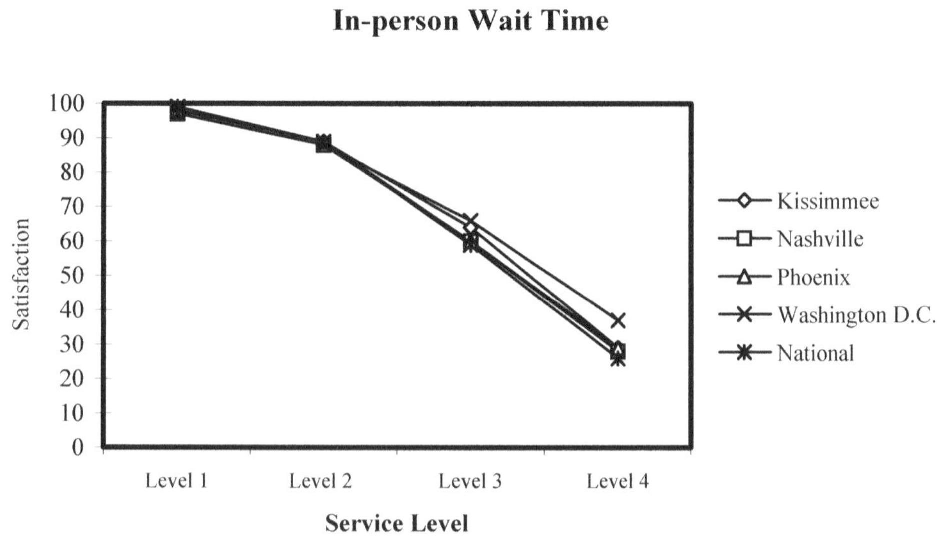

Figure 3.16 Projected satisfaction with customer service in-person wait time in five customer samples

A wait of one minute does not have a serious impact on projected satisfaction. Even at five minutes projected satisfaction is still averaging a respectable 62. Once again, however, it is important to keep in mind that these projected levels of satisfaction reflect an *idea* of waiting a period of time rather than the actual experience. These ratings should be confirmed through further research.

Customer Communications

There was a strong sentiment in the focus groups that water utilities do not communicate effectively with their customers. We also have direct evidence of this in that few focus group participants could recall receiving a water quality report from their utility, suggesting that the attempt to communicate had not been effective. This service category included three attributes: Frequency of General Communication, Schedule of Water Quality Communication, and Internet Information. We had initially intended to include an item measuring preferences for different media of communication but could not fit the question into the format needed to include it in the optimization model.

Frequency of General Communication

Opinion was split about whether utilities should attempt to communicate more frequently than they presently do, some thinking this is imperative, others saying they only want to hear from their water utility when there is a problem. We included one item that addressed all forms of communication, and another that specifically addressed water quality communication. The item for general communication had four levels of service.

Level 1 The utility communicates monthly
Level 2 The utility communicates 4 times/year

Level 3 The utility communicates once/year
Level 4 The utility communicates only in emergencies

The responses from each utility and the national sample are represented in Figure 3.17.

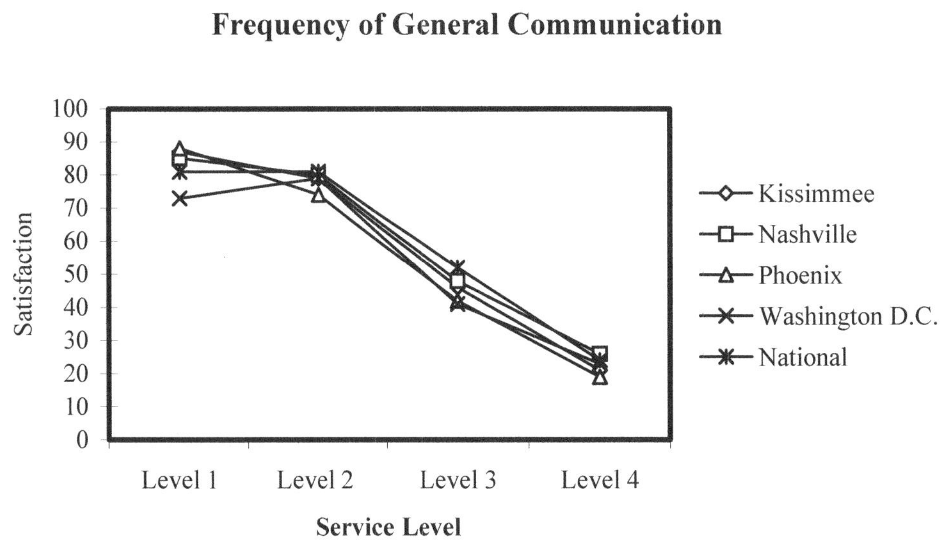

Figure 3.17 Projected satisfaction with frequency of general communication in five customer samples

The grouping of responses from the different samples is not as tight for this attribute as it is for most others, though the shape of the curve is quite similar. It is not altogether surprising that D.C. WASA customers slightly prefer Level 2 service, since this reflects the rate at which their bill arrives. The national sample is indifferent to Levels 1 and 2, probably reflecting a similar pattern of people responding based on the frequency of billing, assuming that most utilities include at least some form of minimal communication with the bill. Even in the samples where people currently are billed monthly, the average drop in satisfaction between monthly and quarterly communication is small, only 9 percentage points. The largest drop in projected satisfaction between Level 1 and 2 comes from Phoenix. We conjecture that this is related to the fact that in Phoenix both the focus groups and survey picked up considerable concern about water taste and quality.

Schedule of Water Quality Information

This attribute specifically addressed water quality information rather than general communication from the water utility. The focus groups revealed that most people were not aware that their utility already distributes an annual water quality report. We looked only at the frequency of communication, not at its content, although clearly content would be an important factor in satisfaction. The item for Water Quality Information had four levels of service.

Level 1 Water quality information is distributed in the mail monthly
Level 2 Water quality information is distributed in the mail every three months

Level 3 Water quality information is distributed in the mail once a year
Level 4 Water quality information is available on the Internet.

The responses from each utility and the national sample are represented in Figure 3.18.

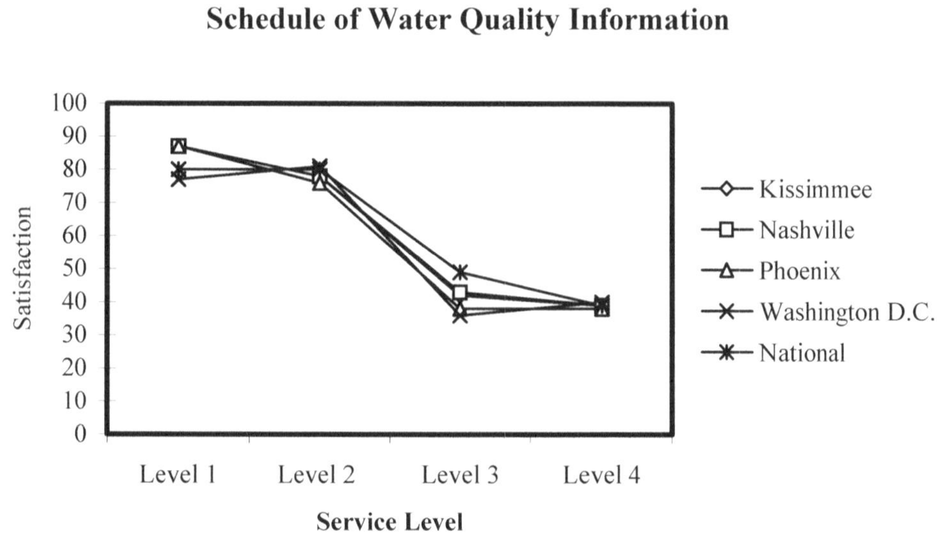

Figure 3.18 Projected satisfaction with the schedule of water quality information in five customer samples

This is another attribute where the pattern of responses differs somewhat from one sample to another. A general tendency is still evident, however. There is relative indifference between Levels 1 and 2 and, at a much lower level of satisfaction, between Levels 3 and 4. The relatively high value for the lower levels of service no doubt reflects, in part, a similar finding of the focus groups, i.e., that some people do not especially care to receive water quality information.

Internet Information

In each focus group session there was at least one person who was comfortable with using the Web to gain information about their water utility. The majority of participants, however, did not seem especially interested. We contemplated dropping this item from the model but were discouraged from doing so by our utility partners, who argued that significant amounts of money are being invested by utilities to establish Web capabilities, and we should test the satisfaction impact of those expenditures.

The item for this attribute had four levels of service reflecting different kinds of information customers could get online from their utilities.

Level 1 Individual account information, bill payment history and general information
Level 2 Individual account information and general information

Level 3 Only general information
Level 4 No Internet information

The responses from each utility and the national sample are represented in Figure 3.19.

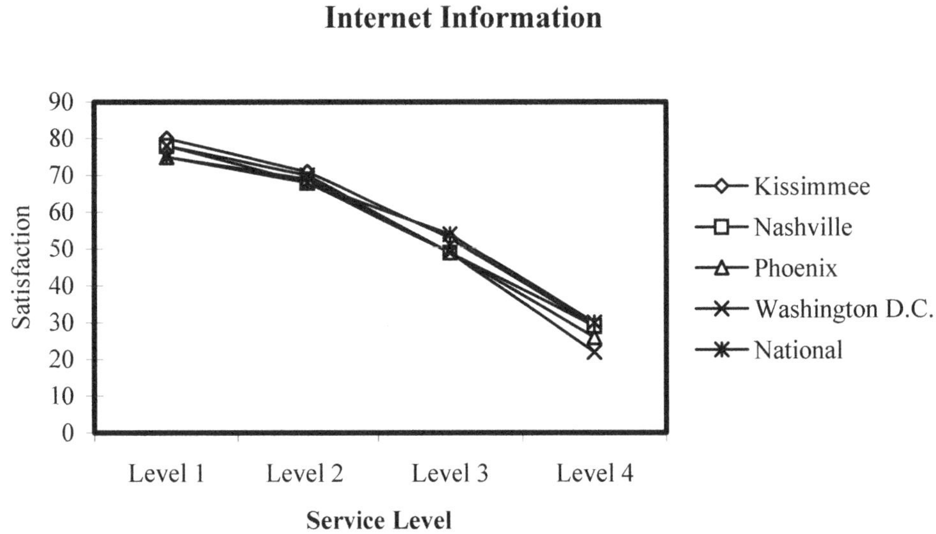

Figure 3.19 Projected satisfaction with Internet information in five customer samples

Unlike the other attributes, where we only averaged the projected satisfaction of people who had actually had an experience with the attribute in question, for this attribute we averaged projected satisfaction of all people who said they merely knew what information their utility distributes. The number of people who reported having logged on to their water utility website was too small (about 4% of the total sample) to serve as the basis of the projections.

RELATIVE IMPORTANCE OF ATTRIBUTES FOR OVERALL SATISFACTION

We have argued that having knowledge of current satisfaction does not provide enough information on which to take actions. The previous section discussed the first of three pieces of additional information that are needed: the rate at which satisfaction rises or falls from any given level of service. This section discusses a second piece of additional information: the relative importance of each attribute in determining overall satisfaction.

We cannot assume that survey respondents have a consistent standard for judging satisfaction across all attributes. If they give a rating of 93 to both 24-hour customer service availability and the resolution of bill problems with no negotiations, we should not assume that these two levels of service would make the same contribution to overall satisfaction. A more likely model is that respondents begin each series of ratings at or near the top of the scale and then work down from there. We need a number—a weight—that we can use to calibrate the attributes against one another with respect to their impact on overall satisfaction. We call this the *importance weight.*

Another kind of weight that is needed is the percentage of customers that is affected by an attribute of service. Changes to some attributes affect nearly everyone. Billing Frequency is a good example. Changes to other attributes affect only small numbers of customers. For instance, only about 4% of customers log on to utility web sites during the course of a year. All other things being equal, changing the level of service for Internet Information cannot have the same impact as changing service for Billing Frequency, because of the different exposure of the customer base to the attributes. We call the weight to account for this difference the *exposure weight*.

Importance Weight

One way to weight the attributes for their importance would be to ask respondents themselves what attributes are most important to them. This approach has some problems, however. It appears that customers do not always have good insight into the question. They know they are satisfied or dissatisfied but do not know which experiences have made them that way. This is especially the case if they have not had any dramatically bad or good experiences. If you ask respondents to rate importance on a scale of 1 to 10, for instance, they will tend to rate all or most attributes close to 10. This is not surprising, since attributes are included in the research because they are expected to be important to customer satisfaction. If you force respondents to make trade-offs, and rank the order of importance, they tend to have a difficult time of it, adding additional burden to the task. In fact, another key reason for choosing not to ask respondents to provide their own attribute importance weights is that it adds considerably to the length of the survey. Also, in ranking the attributes we only learn about the order of their importance. We have no reason to expect that the intervals between ranks will be equal, and this is the very kind information that is needed for creating weights.

The other general approach to weighting is to let the data indicate the importance of individual attributes. This is the approach used for this project. Here, too, however, there are a number of difficult issues.

The most powerful and direct approach would be to use linear regression modeling.[15] Overall satisfaction might be regressed on all of the attributes simultaneously. The regression coefficient for each attribute would represent the unique correlation between that attribute and overall satisfaction, *controlling for all other variables in the model*. This would be a direct estimate of its importance relative to the other attributes. The problem with this approach is that there is not enough variance in the dependent variable, overall satisfaction, nor a large enough sample, to support a model with 17 independent variables, i.e., the 17 attributes. This is true even if the full sample is used to estimate the model. The model "blows up," meaning that many of the coefficients are not statistically significant and take on meaningless values.

For this study, as an alternative to the regression-based approach to estimating attribute weights, we have used Pearson's correlation coefficient r, calculated between each attribute and

---

[15] Linear regression is a statistical technique for finding relationships among variables. A single dependent variable, the value of which we want to predict, is correlated with a set of independent variables, the values of which are believed to be related to the dependent variable. The terminology implies a causal relationship—the independent variables *cause* changes in the dependent variable— but this is really only an assumption of the model. Thus, we assume that overall satisfaction, the dependent variable, is caused by a linear combination of satisfaction with the different attributes of service. For each independent variable the analysis estimates a *coefficient* representing the unique effect of changes in that variable on the dependent variable. The regression coefficients for attribute scores represent the effect of each attribute on overall satisfaction, equivalent to their importance.

overall satisfaction. The size of the coefficient represents the strength of the relationship between the attribute and overall satisfaction. Table 3.3 summarizes importance weights for all attributes and all samples, using this approach. Attributes with particularly high values (0.45 or above) are represented in boldface type, for emphasis.

Table 3.3
Importance and exposure weights for customer service attributes

	Importance weight[1]					Exposure weight[2]				
Attribute	Kissimmee	Nashville	Phoenix	Washington D.C.	National	Kissimmee	Nashville	Phoenix	Washington D.C.	National
Resolve billing problem	**0.47**	**0.58**	**0.52**	**0.50**	**0.47**	0.03	0.18	0.08	0.17	0.08
Billing schedule	0.34	0.35	0.35	**0.47**	0.17	1.00	1.00	1.00	1.00	1.00
Hours of operation: field service	0.30	0.37	0.27	0.38	0.29	0.18	0.15	0.15	0.20	0.20
Appointment window: field service	0.32	0.32	0.28	0.28	0.26	0.18	0.15	0.15	0.20	0.20
Response time for field service	0.34	**0.47**	0.38	0.35	0.41	0.18	0.15	0.15	0.20	0.20
Arrival etiquette: field service	0.40	**0.49**	0.41	0.35	0.40	0.18	0.15	0.15	0.20	0.20
Professionalism: field service	**0.45**	0.40	0.43	0.37	0.35	0.18	0.15	0.15	0.20	0.20
Follow-up: field service	0.37	0.43	0.34	0.42	0.37	0.18	0.15	0.15	0.20	0.20
Hours of operation: customer service	0.28	0.37	0.34	0.39	0.24	0.33	0.43	0.44	0.34	0.29

1 Represents the strength of the relationship (i.e., correlation) between the attribute and overall satisfaction. (1.00 = highest possible importance, 0.00 = lowest possible importance)

2 Represents the percentage of customers that has experienced each attribute within the past 12 months.

(continued)

Table 3.3 (Continued)

Attribute	Importance weight[1]					Exposure weight[2]				
	Kissimmee	Nashville	Phoenix	Washington D.C.	National	Kissimmee	Nashville	Phoenix	Washington D.C.	National
Authority of reps: customer service	0.38	**0.48**	**0.56**	0.42	0.34	0.33	0.43	0.44	0.34	0.29
Professionalism: customer service	0.40	**0.47**	**0.46**	0.35	0.40	0.33	0.43	0.44	0.34	0.29
Automated phone: customer service	0.44	0.39	0.37	0.30	0.33	0.22	0.39	0.30	0.32	0.22
Phone hold time: customer service	**0.45**	**0.46**	0.39	0.33	0.29	0.22	0.39	0.30	0.32	0.22
Line wait: customer service	**0.46**	0.42	0.34	0.37	0.24	0.11	0.16	0.23	0.10	0.13
Frequency utility communicates	0.31	0.34	0.27	0.26	0.34	0.83	0.85	0.90	0.87	0.82
Water quality info schedule	0.30	0.33	0.34	0.27	0.38	0.83	0.85	0.90	0.87	0.82
Internet information	0.33	0.31	0.37	0.11	0.24	0.06	0.02	0.03	0.09	0.02

1 Represents the strength of the relationship (i.e., correlation) between the attribute and overall satisfaction. (1.00 = highest possible importance, 0.00 = lowest possible importance)

2 Represents the percentage of customers that has experienced each attribute within the past 12 months.

The correlation approach to importance weighting has the merit of being simple; however, it does bring with it a problem of its own, the problem of interaction among the attributes. To illustrate, we provide the following example. Note that for the attribute Phone Hold Time, the importance weights for two samples, Kissimmee and Nashville, are quite high, 0.45 and 0.46 respectively. For the other utilities, the weights are middling or low. The lower values are more consistent with the expectations derived from focus groups and the preference contour. A correlation between two variables does not establish that one variable causes the other. Long hold-times in Kissimmee and Nashville, for instance, may not cause dissatisfaction but rather may accompany some other experience that causes dissatisfaction, such as Bill Problems, i.e., the reason customers are calling in the first place.

We tested the possibility that the attribute Resolution of Bill Problems was actually driving overall satisfaction in Kissimmee and Nashville, rather than Hold Time, by regressing Overall Satisfaction on both variables. If the relationship between Hold Time and Overall Satisfaction is spurious, the regression coefficient would be small relative to the coefficient for Bill Problems. The regression coefficient is equivalent to the correlation, but controls for variance explained by other variables in the regression equation, thus removing that influence on the relationship.

It does appear that Resolution of Bill Problem explains some of the relationship between Phone Hold Time and Overall Satisfaction, but not all. The regression coefficient for Phone Hold Time is considerably diminished, but still *significant*—both in the sense of being meaningful and in the sense of being statistically significant. Table 3.4 shows the relationships.

Table 3.4
A comparison of correlation and standardized regression coefficients as estimates of importance for two attributes

Sample	Variable	Correlation Coefficient	Ratio	Regression Coefficient	Ratio
Kissimmee	Resolution of Bill Problem	0.47	1.04	0.39	1.34
	Phone Hold Time	0.45		0.29	
Nashville	Resolution of Bill Problem	0.58	1.26	0.48	2.00
	Phone Hold Time	0.46		0.24	

All statistics are significant, $\alpha<0.05$.

The ratio of the two variables' weights shows their relative importance using the alternative methods of estimation. In each sample, the correlation-based weight is exaggerating the importance of Phone Hold Time relative to the importance of Resolution of Bill Problems. This is seen in the smaller ratio number for the correlation-based weight compared to the regression-based weight. For example, in Kissimmee the two variables have nearly the same importance when correlations are used ($0.47 \div 0.45=1.04$), but Resolution of Bill Problems is substantially more important than Phone Hold Time using regression coefficients ($0.39 \div 0.29=1.34$). This regression model has only two independent variables, however. In the hypothetical model with 17 variables, we could not predict what the relative importance weights of the two variables would be. Unfortunately, such a model cannot be solved with the data at hand. Thus, we use the correlation-based weighting as the best available alternative and try to be alert to possible distortions such as this one.

Exposure Weights

Table 3.4 also shows exposure weights for each attribute and each sample. These represent the percentage of the sample's respondents who report having an experience relevant to a particular attribute within the past 12 months. So, for instance, 3% of the Kissimmee sample and 18% of the Nashville sample said they had a bill problem within the 12 months prior to receiving the survey. We assume that all customers receive a bill, so the exposure weight of Billing Frequency is 1.00. All of the field service attributes reflect the same exposure weight, which represents the percentage of respondents who have had water utility representatives come to their home in the past 12 months.[16] There are three different weights for the Call-in/Walk-in Service category, the first representing the percentage of customers who report either calling or walking into the customer service department, and the latter two reflecting only one or the other form of contact. The exposure weight for Customer Communications reflects the percentage of customers who report seeing or hearing any of seven possible ways utilities might communicate with their customers.[17] The exposure weight for Internet Information reflects the percentage of the sample who say they have logged on to their water utility's website.

There are a number of issues involved in estimating the exposure rate of customers to attributes of customer service. The 12-month time frame is somewhat arbitrary. What is important is to choose a frame within which experiences will be remembered and will contribute directly to overall satisfaction, rather than becoming blurred into an undifferentiated feeling about the utility. It is widely thought that negative experiences are recalled more vividly and for a longer period of time than positive experiences. The 12-month time frame used here does not account for this phenomenon. However, without empirical evidence bearing on the difference in recall, we believe it would be arbitrary to impose different weights. To some extent, any differences in recall will be accounted for in the differential recall of respondents and the tendency to report especially salient experiences as being more recent than they actually were. In other words, the 12-month time frame may actually gather in 18 months of strong experiences and four or six months of weak experiences.

Comparing the exposure weights across utilities does suggest the value that a larger customer service benchmarking study would bring. For instance, the larger percentage of customers contacting Phoenix and Nashville water utilities customer service departments in a given year (44% and 43%, respectively) entails a greater per customer burden of costs than at Kissimmee, where only 33% of customers contact customer service. Better knowledge of these use rates would supplement satisfaction data for customer service decision making.

---

[16] To stimulate recall of such experiences and gather more detailed information, we asked respondents questions about four specific kinds of visits: meter installation or maintenance, leak investigation or repair, service turn-on after disconnection for nonpayment, and meter reading to start or stop service.

[17] The forms of communication include: reading an informational bill insert, reading a water quality report, seeing on TV a public service announcement, hearing on the radio a public service announcement, seeing a billboard, seeing or hearing a news story, or logging onto the Internet.

CUSTOMER SERVICE COSTS

A common sense approach to customer service improvement would be to address areas where satisfaction is lowest, and where improvement would have the greatest impact. But this approach ignores the question of cost. If areas where satisfaction is low are very costly to address, it might be more effective to address two or three alternative areas of service that are cheaper. In other words, information about the costs of providing customer service at different levels provides a constraint against which service can be optimized. When cost and satisfaction data are brought together it is possible to estimate the marginal cost of satisfaction. Then the decision is easy: purchase satisfaction where the cost is least.

We gathered information about costs in a survey of the four partner utilities. In earlier research, we had asked the utility to estimate the cost of providing service at each level of each attribute on the satisfaction survey. This put a significant burden on the utility to estimate exactly what level of effort would have exactly the required effect. That is not a natural way of thinking about service improvements. For this project, we altered the question. We asked only what measures the utility would take if it wanted to improve service on a particular attribute. Then we asked how much those measures would cost and what impact the measures would have on customer satisfaction. Clearly, this last piece of information is subject to error. We promised to do sensitivity testing around the utility estimate, to see how robust the findings would be.

The following tables (Table 3.5 to Table 3.8) present the cost increases and cost savings for changes in service reported by the four partner utilities.[18] We asked the utilities, where possible, to provide two levels of service change in each direction, so we would have cost changes for a total of five service levels: current service, a first improvement, a second improvement, a first economy measure, and a second economy measure. In many instances, the utilities reported there would only be one measure they could take to improve or economize service. In other instances, no changes at all were possible on one or both directions of change. None of this poses a problem for the optimization model. It does, of course, limit the possibilities for improving overall service.

[18] We emphasize that these are incremental costs, i.e., additions and subtractions from current costs.

Table 3.5
Incremental costs for service changes in Kissimmee, Fla.

Attribute	Service change	Measure	Cost impact
Resolve billing problem	1st Increase	Training	$913
	2nd Increase	Training	$1,826
	1st Decrease	None possible	($0)
	2nd Decrease	None possible	($0)
Billing schedule	1st Increase	Hex-a-gram system	$3,500,000
	2nd Increase	None possible	$0
	1st Decrease	None possible	($0)
	2nd Decrease	None possible	($0)
Hours of operation: field service	1st Increase	Increase personnel—2 line techs	$108,824
	2nd Increase	None possible	$0
	1st Decrease	Reduce personnel—2 line techs	($108,824)
	2nd Decrease	Reduce personnel—4 line techs	($217,648)
Appointment window: field service	1st Increase	Cell phones for meter techs	$51,500
	2nd Increase	None possible	$0
	1st Decrease	None possible	($0)
	2nd Decrease	None possible	($0)
Response time for field service	1st Increase	Add staff	$206,410
	2nd Increase	Mobile dispatch	$1,000,000
	1st Decrease	None possible	($0)
	2nd Decrease	None possible	($0)
Arrival etiquette: field service	1st Increase	Cell phones for line techs	$5,580
	2nd Increase	None possible	$0
	1st Decrease	No customer notification of service	($276,480)
	2nd Decrease	None possible	($0)
Professionalism: field service	1st Increase	Training	$913
	2nd Increase	Training	$1,826
	1st Decrease	None possible	($0)
	2nd Decrease	None possible	($0)
Follow-up: field service	1st Increase	Two clerks	$104,000
	2nd Increase	None possible	$0
	1st Decrease	None possible	($0)
	2nd Decrease	None possible	($0)

(continued)

Table 3.5 (Continued)

Attribute	Service change	Measure	Cost impact
Hours of operation: customer service	1st Increase	Add personnel, training	$37,644
	2nd Increase	Add personnel, training	$75,288
	1st Decrease	None possible	($0)
	2nd Decrease	None possible	($0)
Authority of reps: customer service	1st Increase	Add personnel, training	$37,716
	2nd Increase	Add personnel, training	$75,432
	1st Decrease	None possible	($0)
	2nd Decrease	None possible	($0)
Professionalism: customer service	1st Increase	Add personnel, training	$37,716
	2nd Increase	Add personnel, training	$75,432
	1st Decrease	None possible	($0)
	2nd Decrease	None possible	($0)
Automated phone: customer service	1st Increase	No automation, live response	$56,880
	2nd Increase	None possible	$0
	1st Decrease	Complete automation	($56,880)
	2nd Decrease	None possible	($0)
Phone hold time: customer service	1st Increase	Cross training	$12,688
	2nd Increase	Add personnel	$37,500
	1st Decrease	None possible	($0)
	2nd Decrease	None possible	($0)
Line wait: customer service	1st Increase	Add personnel	$37,500
	2nd Increase	None possible	$0
	1st Decrease	None possible	($0)
	2nd Decrease	None possible	($0)
Frequency utility communicates	1st Increase	Quarterly newsletter	$20,000
	2nd Increase	Monthly bill stuffer	$36,000
	1st Decrease	None possible	($0)
	2nd Decrease	None possible	($0)
Water quality info schedule	1st Increase	Quarterly newsletter	$28,000
	2nd Increase	Monthly bill stuffer	$84,000
	1st Decrease	None possible	($0)
	2nd Decrease	None possible	($0)
Internet information	1st Increase	None possible	$0
	2nd Increase	None possible	$0
	1st Decrease	None possible	($0)
	2nd Decrease	None possible	($0)

Table 3.6
Incremental costs for service changes in Nashville, Tenn.

Attribute	Service change	Measure	Cost impact
Resolve billing problem	1st Increase	Automated meter reading	$2,500,000
	2nd Increase	Training	$16,507
	1st Decrease	None possible	($0)
	2nd Decrease	None possible	($0)
Billing schedule	1st Increase	Send bill on same date every month	$120,718
	2nd Increase	EBPP	$2250
	1st Decrease	Bill every two months	($247,526)
	2nd Decrease	None possible	($0)
Hours of operation: field service	1st Increase	Extend weekend hours	$19,218
	2nd Increase	Staff 24/7	$99,507
	1st Decrease	Reduce weekend hours	($19,218)
	2nd Decrease	None possible	($0)
Appointment window: field service	1st Increase	Schedule 4-hour window	$72,455
	2nd Increase	Schedule 2-hour window	$144,910
	1st Decrease	None possible	($0)
	2nd Decrease	None possible	($0)
Response time for field service	1st Increase	Add staff	$206,410
	2nd Increase	Mobile dispatch	$1,000,000
	1st Decrease	None possible	($0)
	2nd Decrease	None possible	($0)
Arrival etiquette: field service	1st Increase	Outbound caller and advanced scheduling	$24,152
	2nd Increase	Complete work on requested date	$1,279,779
	1st Decrease	None possible	($0)
	2nd Decrease	None possible	($0)
Professionalism: field service	1st Increase	Training	$28,796
	2nd Increase	Brochures and website	$15,094
	1st Decrease	None possible	($0)
	2nd Decrease	None possible	($0)
Follow-up: field service	1st Increase	Door hangers	$5,616
	2nd Increase	Personal contact from leads	$25,081
	1st Decrease	None possible	($0)
	2nd Decrease	None possible	($0)

(continued)

Table 3.6 (Continued)

Attribute	Service change	Measure	Cost impact
Hours of operation: customer service	1st Increase	Advertise operating hours	$10,000
	2nd Increase	Add staff and answering service	$25,000
	1st Decrease	Reduce business hours	($120,000)
	2nd Decrease	Eliminate lobby service	($108,000)
Authority of reps: customer service	1st Increase	Training	$17,000
	2nd Increase	Eliminate answering service, add reps	$12,000
	1st Decrease	None possible	($0)
	2nd Decrease	None possible	($0)
Professionalism: customer service	1st Increase	Training	$30,000
	2nd Increase	Call monitoring equipment	$60,000
	1st Decrease	Reduce training and reps	($66,000)
	2nd Decrease	None possible	($0)
Automated phone: customer service	1st Increase	Send customers brochure on IVR	$10,000
	2nd Increase	Give customer option to speak to rep sooner	$10,000
	1st Decrease	None possible	($0)
	2nd Decrease	None possible	($0)
Phone hold time: customer service	1st Increase	Add 3 service reps, extend hours	$180,000
	2nd Increase	24-hour service with reps.	$120,000
	1st Decrease	Eliminate 3 service reps.	($108,000)
	2nd Decrease	Eliminate 5 reps	($180,000)
Line wait: customer service	1st Increase	Add 1 service rep in lobby	$36,000
	2nd Increase	Add kiosk and 2 additional reps	$82,000
	1st Decrease	Eliminate 2 service reps.	($72,000)
	2nd Decrease	Eliminate lobby, add phones	($108,000)
Frequency utility communicates	1st Increase	None possible	$0
	2nd Increase	None possible	$0
	1st Decrease	None possible	($0)
	2nd Decrease	None possible	($0)
Water quality info schedule	1st Increase	None possible	$0
	2nd Increase	None possible	$0
	1st Decrease	None possible	($0)
	2nd Decrease	None possible	($0)
Internet information	1st Increase	Interactive website	$10,000
	2nd Increase	Interactive website	$5014
	1st Decrease	None possible	($0)
	2nd Decrease	None possible	($0)

Table 3.7
Incremental costs for service changes in Phoenix, Ariz.

Attribute	Service change	Measure	Cost impact
Resolve billing problem	1st Increase	Bill-problem tracking system	$70,000
	2nd Increase	None possible	$0
	1st Decrease	None possible	($0)
	2nd Decrease	None possible	($0)
Billing schedule	1st Increase	None possible	$0
	2nd Increase	None possible	$0
	1st Decrease	None possible	($0)
	2nd Decrease	None possible	($0)
Hours of operation: field service	1st Increase	One additional evening hour	$209,617
	2nd Increase	One additional morning hour	$209,617
	1st Decrease	One fewer morning hours	($209,617)
	2nd Decrease	One fewer evening hours	($209,617)
Appointment window: field service	1st Increase	None possible	$0
	2nd Increase	None possible	$0
	1st Decrease	None possible	($0)
	2nd Decrease	None possible	($0)
Response time for field service	1st Increase	New system and additional staff	$2,636,854
	2nd Increase	None possible	$0
	1st Decrease	None possible	($0)
	2nd Decrease	None possible	($0)
Arrival etiquette: field service	1st Increase	Develop new procedure	$10,800
	2nd Increase	None possible	$0
	1st Decrease	Reduce staff	($209,617)
	2nd Decrease	None possible	($0)
Professionalism: field service	1st Increase	Training	$50,000
	2nd Increase	Training	$25,000
	1st Decrease	None possible	($0)
	2nd Decrease	None possible	($0)
Follow-up: field service	1st Increase	Tracking system to ensure follow-up	$50,000
	2nd Increase	Follow-up phone call with a letter	$414,110
	1st Decrease	Leave door card only	($41,923)
	2nd Decrease	None possible	($0)

(continued)

Table 3.7 (Continued)

Attribute	Service change	Measure	Cost impact
Hours of operation: customer service	1st Increase	Add two hours per day	$513,864
	2nd Increase	Add Saturday hours	$742,330
	1st Decrease	Close one pay station	($75,000)
	2nd Decrease	None possible	($0)
Authority of reps: customer service	1st Increase	None possible	$0
	2nd Increase	None possible	$0
	1st Decrease	None possible	($0)
	2nd Decrease	None possible	($0)
Professionalism: customer service	1st Increase	Additional staff and training	$362,576
	2nd Increase	None possible	$0
	1st Decrease	None possible	($0)
	2nd Decrease	None possible	($0)
Automated phone: customer service	1st Increase	None possible	$0
	2nd Increase	None possible	$0
	1st Decrease	None possible	($0)
	2nd Decrease	None possible	($0)
Phone hold time: customer service	1st Increase	Add trunk lines, staff, automation	$356,932
	2nd Increase	Add staff, automation	$185,000
	1st Decrease	None possible	($0)
	2nd Decrease	None possible	($0)
Line wait: customer service	1st Increase	None possible	$0
	2nd Increase	None possible	$0
	1st Decrease	None possible	($0)
	2nd Decrease	None possible	($0)
Frequency utility communicates	1st Increase	None possible	$0
	2nd Increase	None possible	$0
	1st Decrease	None possible	($0)
	2nd Decrease	None possible	($0)
Water quality info schedule	1st Increase	None possible	$0
	2nd Increase	None possible	$0
	1st Decrease	Bi-monthly newsletter instead of monthly	($150,000)
	2nd Decrease	None possible	($0)
Internet information	1st Increase	Add Internet bill payments	$150,000
	2nd Increase	Add account information	$85,644
	1st Decrease	None possible	($0)
	2nd Decrease	None possible	($0)

Table 3.8
Incremental costs for service changes in Washington, D.C.

Attribute	Service change	Measure	Cost impact
Resolve billing problem	1st Increase	Training, automated bill processing	$160,000
	2nd Increase	Raise salaries	$120,000
	1st Decrease	Reduce staff and training	($170,000)
	2nd Decrease	Reduce salaries	($100,000)
Billing schedule	1st Increase	Customer billing-cycle selection	$200,000
	2nd Increase	None possible	$0
	1st Decrease	Eliminate monthly billing	($50,000)
	2nd Decrease	Bring billing back in-house	($1,000,000)
Hours of operation: field service	1st Increase	Extend hours—12/7	$400,000
	2nd Increase	Extend hours—24/7	$800,000
	1st Decrease	Reduce hours	($300,000)
	2nd Decrease	Reduce hours	($500,000)
Appointment window: field service	1st Increase	Automated dispatch system with GPS—2-hour window	$1,900,000
	2nd Increase	Specific time appointments	$200,000
	1st Decrease	Reduce staff—6-hour window	($1,000,000)
	2nd Decrease	Only day specified	($300,000)
Response time for field service	1st Increase	Automated dispatch system with GPS—2-hour window	$1,900,000
	2nd Increase	Increase training, salaries	$300,000
	1st Decrease	Reduce staff	($1,000,000)
	2nd Decrease	Decrease salaries	($200,000)
Arrival etiquette: field service	1st Increase	Equipment, staff, training	$1,070,000
	2nd Increase	Increase training, salaries	$200,000
	1st Decrease	Decrease staff	($350,000)
	2nd Decrease	Decrease salaries	($100,000)
Professionalism: field service	1st Increase	Additional training, information access, performance review	$145,000
	2nd Increase	Increase salaries	$200,000
	1st Decrease	Decrease training	($50,000)
	2nd Decrease	Decrease salaries	($100,000)
Follow-up: field service	1st Increase	Increase staff and training, outsource	$250,000
	2nd Increase	Automated follow-up	$300,000
	1st Decrease	Decrease training	($50,000)
	2nd Decrease	None possible	($0)

(continued)

Table 3.8 (Continued)

Attribute	Service change	Measure	Cost impact
Hours of operation: customer service	1st Increase	24/7 call center	$600,000
	2nd Increase	24/7 web monitoring, increase awareness of IVR	$120,000
	1st Decrease	Reduce service hours	($100,000)
	2nd Decrease	None possible	($0)
Authority of reps: customer service	1st Increase	Training, reorganization	$45,000
	2nd Increase	Cross-functional training, increase salaries	$200,000
	1st Decrease	Shift services	($50,000)
	2nd Decrease	Decrease salaries	($100,000)
Professionalism: customer service	1st Increase	Increase training, monitoring and feedback	$150,000
	2nd Increase	Increase salaries	$200,000
	1st Decrease	Reduce training and monitoring	($40,000)
	2nd Decrease	Decrease salaries	($100,000)
Automated phone: customer service	1st Increase	Increase staff, training	$175,000
	2nd Increase	Increase salaries	$200,000
	1st Decrease	Decrease staff, training	$(170,000)
	2nd Decrease	Decrease salaries	($100,000)
Phone hold time: customer service	1st Increase	Increase staff, IVR functionality, customer communication	$250,000
	2nd Increase	Increase salaries, training	$220,000
	1st Decrease	Decrease staff, training	$(170,000)
	2nd Decrease	Decrease salaries	($100,000)
Line wait: customer service	1st Increase	Increase staff, greeter, kiosk, system functionality	$315,000
	2nd Increase	Increase salaries, training	$120,000
	1st Decrease	Decrease staff	$(150,000)
	2nd Decrease	Decrease salaries, training	($110,000)
Frequency utility communicates	1st Increase	Formatting changes, cable announcements	$100,000
	2nd Increase	Newspaper and radio announcements	$110,000
	1st Decrease	Phone recording and web message	($70,000)
	2nd Decrease	Bill message, in-house printing	($45,000)
Water quality info schedule	1st Increase	Added Web, newspaper, bill messages	$95,000
	2nd Increase	Increase special mailings	$30,000
	1st Decrease	Bill message/ web page only	($135,000)
	2nd Decrease	Telephone recording	($45,000)
Internet information	1st Increase	Last 3 payments information	$100,000
	2nd Increase	Discount for Web payment	$150,000
	1st Decrease	None possible	($0)
	2nd Decrease	None possible	($0)

OPTIMAL LEVELS OF SERVICE

The previous findings provide information needed for the optimization of customer service offerings. We have estimates of how satisfied customers currently are, how satisfied they would be at higher and lower levels of service, and how much it would cost to provide service at those levels. The optimal solution is obtained by calculating the cost per unit satisfaction of each level of each attribute and reassigning dollars and people to levels of service in order to maximize satisfaction within a budget constraint. It might be possible to come close to an optimum by trial and error. That is the approach that many of the so-called "simulators" take. This is not the most straightforward approach, however, as there are 28,625 combinations of attributes that could be tried. We have treated the solution as a linear programming problem (see Chvatal 1983, Gass 1970).

There is more than one way to model an optimal solution using linear programming. For this research we have used the model where the objective function (the function to be maximized) is the sum of the products of the percentage of customers at each level of each attribute and the weighted satisfaction derived at each level. The function is presented in Chapter 2 (see page 23).

The value of the cost constraint function is set equal to zero in order to make changes in service neutral with respect to cost. That is, the solution maximizes satisfaction at current costs. The value of the cost function could also be set higher or lower, to indicate where additional funds should be added, or where cost reductions should be undertaken.

Table 3.9 presents the results of our modeling of the four partner utilities' data. For each attribute, we present the estimated satisfaction derived from each level of service, and the cost of providing service at that level. It should be noted that the satisfaction values represented in the table are unweighted, to preserve their common sense meaning.

The estimated optimal level of service is highlighted in gray. For example, the estimated optimal level of service for the attribute "Resolution of Bill Problems" is an increase of two levels for Kissimmee and Washington D.C., an increase of one level for Phoenix, and the current level of service for Nashville. This would earn Kissimmee 6 "satisfaction points" (i.e., 85 – 79 = 6) at a cost of $2,000; Washington D.C. would get 13 satisfaction points at a cost of $280,000, and so on.

It is easy to see how the program arrives at an optimum. A unit cost of satisfaction is calculated for each level of each attribute. For instance, in the example above, the cost of a satisfaction point for Kissimmee raising service on "Bill Problem Resolution" 2 levels is $2,000 ÷ 6 = $333. (The weighting of attributes makes the satisfaction points represented in this table not directly comparable.) The program finds the unique combination of levels that will maximize satisfaction while keeping costs constant.

Table 3.9
Optimal service levels at four utilities

Attribute	Utility	Satisfaction (unweighted)					Incremental costs ($1,000)				
		+2	+1	0	-1	-2	+2	+1	0	-1	-2
Resolve billing problem	Kissimmee	**85**	82	79			**2**	1	0		
	Nashville	87	86	**81**			2517	2500	**0**		
	Phoenix		**87**	82				**70**	0		
	Washington D.C.	**81**	75	68	50	31	**280**	160	0	-170	-270
Billing schedule	Kissimmee		**93**	91				**4**	0		
	Nashville	91	91	**91**	68		123	121	**0**	-248	
	Phoenix			**95**					**0**		
	Washington D.C.			**68**					**0**		
Hours of operation: field service	Kissimmee		74	72	65	**55**		109	0	-109	**-218**
	Nashville	80	75	71	**68**		100	19	0	**-19**	
	Phoenix	74	72	71	63	**61**	419	210	0	-210	**-419**
	Washington D.C.	76	69	68	**58**	44	800	400	0	**-300**	-500
Appointment window: field service	Kissimmee		77	**74**				52	**0**		
	Nashville	82	76	**70**			145	72	**0**		
	Phoenix			**67**					**0**		
	Washington D.C.	90	70	63	**44**	20	2100	1900	0	**-1000**	-1300
Response time for field service	Kissimmee	91	**87**	71			99	**49**	0		
	Nashville	81	76	**71**			1206	206	**0**		
	Phoenix		79	**70**				2637	**0**		
	Washington D.C.	87	82	61	**42**	27	2200	1900	0	**-1000**	-1200
Arrival etiquette: field service	Kissimmee		89	**83**	49			6	**0**	-276	
	Nashville	91	**86**	80			1304	**24**	0		
	Phoenix		**87**	84				**11**	0	-210	
	Washington D.C.	97	92	**75**	45	26	1270	1070	**0**	-350	-450
Professionalism: field service	Kissimmee	93	90	**83**			2	1	**0**		
	Nashville	88	83	**81**			44	29	**0**		
	Phoenix	**90**	86	82			**75**	50	0		
	Washington D.C.	95	**87**	73	43	27	345	**145**	0	-50	-150
Follow-up: field service	Kissimmee		88	**80**				104	**0**		
	Nashville	81	**78**	75			25	**6**	0		
	Phoenix	81	**78**	73	67		464	**50**	0	-42	
	Washington D.C.	88	**82**	66	62		550	**250**	0	-50	

(continued)

Table 3.9 (Continued)

		Satisfaction (unweighted)					Incremental costs ($1,000)				
Attribute	**Utility**	**+2**	**+1**	**0**	**-1**	**-2**	**+2**	**+1**	**0**	**-1**	**-2**
Hours of operation: customer service	Kissimmee		91	68				38	0		
	Nashville	84	78	71	66	60	35	10	0	-120	-108
	Phoenix	77	71	69	66		1256	514	0	-75	
	Washington D.C.		91	66	62		720	600	0	-100	
Authority of reps: customer service	Kissimmee		95	86				38	0		
	Nashville	90	89	80			29	17	0		
	Phoenix			84					0		
	Washington D.C.	94	92	76	66	51	245	45	0	-50	-150
Professionalism: customer service	Kissimmee		90	86				38	0		
	Nashville	90	89	82	66	51	90	30	0	-66	-126
	Phoenix		87	82				363	0		
	Washington D.C.	85	78	69	61	48	350	150	0	-40	-140
Automated phone: customer service	Kissimmee		80	72	54			57	0	-57	
	Nashville	79	74	62			20	10	0		
	Phoenix			64					0		
	Washington D.C.	81	72	55	39	24	375	175	0	-170	-270
Phone hold time: customer service	Kissimmee	96	87	79			50	13	0		
	Nashville	90	83	72	66	58	180	180	0	-108	-288
	Phoenix	87	83	76			542	357	0		
	Washington D.C.	94	76	61	37	32	470	250	0	-170	-270
Line wait: customer service	Kissimmee		92	80				38	0		
	Nashville	84	79	76	69	54	82	36	0	-72	-180
	Phoenix			73					0		
	Washington D.C.	94	80	64	49	40	435	315	0	-150	-260
Frequency utility communicates	Kissimmee	82	81	74			36	20	0		
	Nashville			72					0		
	Phoenix			71					0		
	Washington D.C.	79	78	67	55	39	210	100	0	-70	-115
Water quality info schedule	Kissimmee	83	77	68			84	28	0		
	Nashville			68					0		
	Phoenix	68	68	68	59				0	-150	
	Washington D.C.	78	77	66	69	45	125	95	0	-135	-180

(continued)

Table 3.9 (Continued)

		Satisfaction (unweighted)					Incremental Costs ($1,000)				
Attribute	**Utility**	+2	+1	0	-1	-2	+2	+1	0	-1	-2
Internet information	Kissimmee			69					0		
	Nashville	68	65	61			15	10	0		
	Phoenix	76	71	66			236	150	0		
	Washington D.C.	74	70	62			250	100	0		

The model recommendations vary substantially from one utility to another, reflecting the complex relationship among the variables. There are certain general patterns, however. For instance, three of the four partner utilities have recommendations to increase service for Resolve Billing Problems. The fourth utility, Nashville, estimated a disproportionately large cost for increasing service the first step.

All of the utilities get money to pay for service increases by decreasing service to the attribute Hours of Operation for field service. In the case of Kissimmee, which offered few options for cost savings, this was the only service reduction recommended, achieving $218,000 in savings that is distributed over nine other attributes for service improvements. A similar source of recommended revenue for Nashville Metro and D.C. WASA is Line Wait for walk-in service.

Three of the four utilities get a recommendation to increase service for the attributes Follow-Up for field service, Hours of Operation for walk-in/call-in service, Authority of Representatives for walk-in/call-in service, Professionalism for walk-in/call-in service, and Phone Hold Times for walk-in/call-in service. Note that none of these attributes has the lowest current satisfaction ratings. In fact, a number of them are in the mid 80's, which is quite high. But these are attributes where increases in service either bring relatively large increases in satisfaction, cost relatively small amounts of money, or both.

A Sensitivity Test

The reader will recall that one feature of the cost exercise is an estimate, made by the utility, of the impact of changes in service on customer satisfaction. For instance, the customer service manager might estimate that lengthening the hours of service will raise the satisfaction of one-third of customers. It would not be 100% of customers because many would not notice the change, and others would not directly benefit from it. An easy way to treat this one-third change is to say that one-third of customers at each level of perceived service would move to the next higher level of service.

In the future, hard data may be available on which to base these estimates. In the absence of such data, however, the expertise of utility personnel is the best resource available. A question arises, then, about the sensitivity of the results to the accuracy of this estimate. If the recommendations of the model change dramatically with small changes in the estimate, the model is *sensitive* to the estimate. Very sensitive models may require information with a degree of accuracy that is impossible for managers to obtain and would therefore not be useful. For instance, if changing the estimate of change impacts by a small percentage would have a large effect on the recommended changes in service the recommendations would not be trustworthy.

If, on the other hand, the recommendations remain basically the same, even if the change impact estimates are off by a significant amount, the findings are robust or insensitive to the estimate.

One way to test the sensitivity of the current model is to take a more pessimistic view than was taken by our partner utility informants. Perhaps a process of wishful thinking tends to cause managers to exaggerate the positive impact of service improvements, and minimize the impact of service reductions. In any case, we can test the effect on the model's recommendations if such a bias *were* at work, by systematically reducing the estimated impact of service improvements by 50%, and increasing the estimated impact of service reductions by 50%. So if a service increase was originally estimated to affect 50% of customers the model is run assuming an affect on 25% of customers; if a service decrease was originally estimated to affect 60% of customers the model is run assuming an effect on 90% of customers.[19] If the model is not sensitive to such a bias, its recommendations will not alter appreciably.

Of course, this test changes all attributes by the same amount, so it might be argued that the relations among the attributes would remain relatively constant. A different test of the model is to see how it reacts when only some of the estimates are biased. A useful test is to consider only the most radical recommendations for revision, discounting by 50% only those attributes where the original model recommended a two-level increase or decrease in service.

Table 3.10 shows both kinds of sensitivity test carried out with D.C. WASA data.[20] For each attribute a recommendation is given (represented by the shaded square) for each of three scenarios: the baseline, using the utility's estimates; a 1st pessimistic scenario, where all estimates are systematically biased; and a 2nd pessimistic scenario, where only the 2-level change recommendations in the baseline are biased.

[19] These strike us as rather substantial margins of error.

[20] We used D.C. WASA data because they provided the most options for offering different levels of service. This allows more room for the model to give aberrant recommendations under different assumptions.

Table 3.10
Sensitivity of service recommendations to assumptions about satisfaction changes—Washington, D.C.

Attribute	Assumption	Recommendation				
Resolve billing problem	Base	+2	+1	0	–1	–2
	Pessimistic 1	+2	+1	0	–1	–2
	Pessimistic 2	+2	+1	0	–1	–2
Billing schedule	Base	+2	+1	0	–1	–2
	Pessimistic 1	+2	+1	0	–1	–2
	Pessimistic 2	+2	+1	0	–1	–2
Hours of operation: field service	Base	+2	+1	0	–1	–2
	Pessimistic 1	+2	+1	0	–1	–2
	Pessimistic 2	+2	+1	0	–1	–2
Appointment window: field service	Base	+2	+1	0	–1	–2
	Pessimistic 1	+2	+1	0	–1	–2
	Pessimistic 2	+2	+1	0	–1	–2
Response time for field service	Base	+2	+1	0	–1	–2
	Pessimistic 1	+2	+1	0	–1	–2
	Pessimistic 2	+2	+1	0	–1	–2
Arrival etiquette: field service	Base	+2	+1	0	–1	–2
	Pessimistic 1	+2	+1	0	–1	–2
	Pessimistic 2	+2	+1	0	–1	–2
Professionalism: field service	Base	+2	+1	0	–1	–2
	Pessimistic 1	+2	+1	0	–1	–2
	Pessimistic 2	+2	+1	0	–1	–2
Follow-up: field service	Base	+2	+1	0	–1	–2
	Pessimistic 1	+2	+1	0	–1	–2
	Pessimistic 2	+2	+1	0	–1	–2
Hours of operation: customer service	Base	+2	+1	0	–1	–2
	Pessimistic 1	+2	+1	0	–1	–2
	Pessimistic 2	+2	+1	0	–1	–2
Authority of reps: customer service	Base	+2	+1	0	–1	–2
	Pessimistic 1	+2	+1	0	–1	–2
	Pessimistic 2	+2	+1	0	–1	–2

(continued)

Table 3.10 (Continued)

Attribute	Assumption	Recommendation				
Automated phone: customer service	Base	+2	+1	0	–1	–2
	Pessimistic 1	+2	+1	0	–1	–2
	Pessimistic 2	+2	+1	0	–1	–2
Phone hold time: customer service	Base	+2	+1	0	–1	–2
	Pessimistic 1	+2	+1	0	–1	–2
	Pessimistic 2	+2	+1	0	–1	–2
Line wait: customer service	Base	+2	+1	0	–1	–2
	Pessimistic 1	+2	+1	0	–1	–2
	Pessimistic 2	+2	+1	0	–1	–2
Frequency utility communicates	Base	+2	+1	0	–1	–2
	Pessimistic 1	+2	+1	0	–1	–2
	Pessimistic 2	+2	+1	0	–1	–2
Water quality info schedule	Base	+2	+1	0	–1	–2
	Pessimistic 1	+2	+1	0	–1	–2
	Pessimistic 2	+2	+1	0	–1	–2
Internet information	Base	+2	+1	0	–1	–2
	Pessimistic 1	+2	+1	0	–1	–2
	Pessimistic 2	+2	+1	0	–1	–2

It is encouraging that none of the recommendations of the baseline are reversed by the alternative scenarios. Of the 17 attributes, eight do not change under any of the alternatives. Among the nine remaining attributes, only two involve a two-level change in service. Comparing the first pessimistic scenario to the baseline scenario, eight recommendations are different. Half of these, however, entail different levels of change in the same direction. The other half entail recommendations of service changes in one scenario where no change is recommended in the other. Comparing the second pessimistic scenario to the baseline scenario we see that the recommendations are closer. Of the seventeen attributes, only two recommendations change.

We believe these findings indicate that the models are reasonably robust with respect to the estimates of change impacts. Nevertheless, sensitivity testing using a variety of scenarios for key decision variables would be an indispensable part of any application of this method, prior to basing decisions on the findings.

The Impact of Service Changes on Overall Satisfaction

An important question to ask, once the modeling results are calculated, is how much of an improvement in customer satisfaction the recommended changes will make. The linear programming solution calculates an objective value, which is the numerical representation of satisfaction, weighted by importance and exposure. We can remove the weights and compare this value to the estimated satisfaction derived from current service, and get an idea of the predicted increase. Table 3.11 presents the findings for the four partner utilities.

Table 3.11
Estimates of current and optimal satisfaction given constant costs

	Current satisfaction	Optimal satisfaction	Percentage increase
Kissimmee	79	85	8%
Nashville	79	80	1%
Phoenix	77	78	2%
Washington, D.C.	67	78	17%

It is worth noting that the current overall satisfaction values arrived at through the model are lower than the values arrived at either through the single survey item or the common factor analysis (see Table 3.1). It is encouraging that the order of the utilities from highest to lowest is consistent, and that the intervals between them are close to the survey item intervals, with the exception that the Washington sample lags ten points behind Phoenix rather than about four. It is not surprising that these numbers should be different because they have been built up in an entirely different way, summing weighted attribute satisfaction scores rather than category satisfaction scores. It is possible that the scores built up from attributes are lower because respondents are somewhat more critical when they focus on details than they are when they think of the utility as a whole.

These findings may at first be disappointing for Nashville Metro and Phoenix Water Services Department, which are predicted to improve customer satisfaction by only 1 or 2 percent. On reflection, it is not altogether surprising that these two utilities improve so little under the base set of assumptions. The first thing to note is that both of these utilities are already rated rather highly by their customers, so there is relatively little headroom compared to D.C. WASA. An equally important point, however, is that the Nashville and Phoenix utilities provided relatively few options for reducing service, and thus saving costs. Assuming fixed costs, the model can only increase service in one area by reducing service in another. With few options for reduction there are few options for improvement. D.C. WASA provided a very full complement of reduction and improvement options, and thus had more flexibility in reassigning costs.

Investing New Resources in Customer Service

A different way to run the optimization model is to calculate optimal service levels for an additional investment of capital. This approach should open up Nashville and Phoenix to more opportunities for satisfaction improvements because there will be options for improving service without decreasing service in other areas. Table 3.12 presents estimated optimal satisfaction increases (over current satisfaction) for the no-cost model and for five models entailing increasingly larger investments.

Table 3.12
The effect of customer service investments on optimal customer satisfaction

	Current costs	Invest $200,000	Invest $500,000	Invest $1,000,000	Invest $2,000,000	Invest $5,000,000
Kissimmee	8%	9%	10%	10%	10%	10%
Nashville	1%	2%	4%	4%	5%	5%
Phoenix	1%	1%	2%	2%	3%	3%
Washington, D.C.	17%	17%	18%	19%	21%	24%

For D.C. WASA the model predicts a 7 percentage point increase in customer satisfaction with an investment of $5 million in customer service (an astounding 24 percentage point above current satisfaction). Nashville Metro only increases four percentage points above the no-cost model (5 percentage point overall increase) with a $5 million investment. (It should be noted that a 4 or 5 percentage-point increase in customer satisfaction would still constitute a respectable gain.) Phoenix Water Services Department and Kissimmee Department of Water Resources only increase satisfaction two percentage points above the no-cost scenario.

In the case of Nashville, the $5 million investment brings it close to the upper limit of satisfaction that the model can predict. All attributes but one, Resolve Billing Problems, are at their highest level. The Phoenix satisfaction scores increase more slowly for a different reason, however. There, the cost of service improvements is greater, so the $5 million investment has less impact.

CHAPTER 4
LIMITATIONS AND APPLICATIONS OF THE METHOD

This chapter attempts to put the method developed over the previous two chapters into a broader context. We step back to consider the various limitations of the model and point toward possible remedies. We also discuss various partial applications of the method, for utilities that are convinced of the value of customer satisfaction research but lack the resources to fully implement what has been described here.

LIMITATIONS OF THE MODEL

Although the method developed in this report represents a substantial advance over the point-estimate approach commonplace in customer satisfaction research, it has a variety of limitations. Above all, the timeworn adage for model builders applies here no less than anywhere else: "Garbage in; garbage out." That is, the quality of the recommendations is completely dependent on the quality of information going into the analysis. In a number of ways, it is possible to anticipate where error will enter in. It is crucial to be aware of these potential sources of error so that their effect can be minimized, and so that recommendations emerging from the analysis will be given their proper weight, i.e., neither magical pronouncement from an omniscient authority, nor useless random guesses, but somewhere in between: i.e., rational estimates based on the best available information.

In the following sections, the principal limitations of the model are discussed. In all cases we believe the error introduced by these characteristics can be remedied through careful analysis and a variety of countermeasures.

Accuracy of Satisfaction Estimates

The estimation of customer satisfaction profiles is based on the assumption that customers can predict their satisfaction with levels of service they have not yet received from the utility. As discussed in the treatment of phone hold times and line wait times, these projections of satisfaction are subject to error. It is important to understand the nature of this error so that its significance is not exaggerated.

Satisfaction at each level of service is a mean value calculated over the entire sample of replies. It is not important that an individual misjudges his or her satisfaction at a given level of service as long as the misjudgments of all respondents are randomly distributed around the mean value. In other words, the averaging greatly improves our confidence in the final value insofar as we can assume that one person's overestimation of satisfaction is counteracted by another person's underestimation.[21] A problem occurs only when there is systematic bias, that is, where there is a prevailing tendency to misjudge in one direction or the other. It seems likely that many of these systematic biases, such as the wait-time bias we hypothesized, can be anticipated and factored into the interpretation of results.

It is also important to recognize that respondents typically are not projecting satisfaction on the basis of a complete lack of information. To the contrary, most attributes are for aspects of

[21] We believe the dramatic consistency of mean estimates across the samples drawn for this research is another indication of the power of the method to capture an underlying reality.

service that could be received from any number of organizations, and it is safe to assume most respondents have experienced a range of service levels on the majority of them.

Accuracy of the Weights

The attributes of service are weighted in the model to reflect their importance in determining overall satisfaction and the exposure of customers to them. This was discussed in Chapter 3. Two alternative approaches to importance weighting were considered, using standardized regression coefficients or correlation coefficients, but we concluded that only the correlation coefficients represent a realistic option. We showed, however, that correlation-based weights could exaggerate the importance of certain attributes that share variance with other attributes.

Even if regression is not the ultimate basis of weighting the attributes, it could be used to identify shared variance problems by analyzing models with subsets of variables that are suspected of sharing variance with overall satisfaction. That was the approach taken to demonstrate the problem in Chapter 3. From this analysis could come a basis for discounting correlation weights that were clearly overstating the importance of an attribute.

A weighting procedure based on factor analysis could possibly be developed that would combine the best elements of both regression-based and correlation-based approaches. We anticipate further testing of this approach using data from the four partner utilities.

Accuracy of the Change Impact Estimates

When changes in service are made they do not affect everyone. Some people do not have contact with the company in the area of change. Changing the hours when field service is available has no impact on the many customers who do not need field service. Other customers are affected by the attribute, but in a way that is irrelevant to the change. For instance, extending the hours of field service may have no impact on people who need service within the former hours of service. Others may encounter a new level of service but not notice that anything has changed. The change impact is the percentage of people who are affected by the new level of service *and who respond at a higher level of satisfaction than they otherwise would have.*

For this research, we did not have direct measures of these change impacts, so we asked personnel from the partner utilities to estimate their magnitude. We conjecture that utility estimates will exaggerate the true impact of changes, possibly overestimating the effect of increases in service and underestimating the effect of reductions in service. We would not, however, assume that this pattern would always be seen, and might even reverse itself in some circumstances.

In Chapter 3, a sensitivity test of the change impact estimates was conducted for D.C. WASA and it was concluded that the model's findings were reasonably robust. Nevertheless, the need to make such estimates is a limitation of the model at the present time. We believe it would be possible to develop research to gather empirical evidence on the impact of changes in service. It would require cooperation on the part of a large number of utilities, however, and possibly entail risks (i.e., reductions in service) that utilities would be reluctant to take. In the near term the refinement of expert judgments and sensitivity testing seem the best approaches to deal with uncertainty surrounding change impacts

Interactions Among Attributes

The model treats each service change as if its effects were entirely isolated from other attributes, with no interaction effects. Thus, increasing the professionalism of customer service representatives is assumed to be distinct from decreasing the wait for service. When service changes are contemplated, however, it becomes evident that to some extent these attributes are related. Inasmuch as the attribute professionalism is addressed by hiring additional personnel, so that representatives will have more time to spend with each customer, the wait in line may also be reduced. For modeling purposes we assumed that additional personnel would use up all the extra labor time in attending to each customer; but the reality is that the added labor may well be split between additional time per customer and shorter wait times.

For another example of an interaction between attributes, the improvement of bill problem resolution would reduce the number of calls to the customer service center and thus improve phone hold times and in-person line waits. Of course, a utility could also choose to economize by reducing customer service staff. The model in its present form does not account for these effects.

Without interactions built into the model there will be a tendency for it to underestimate both increases in satisfaction from improvements and decreases in satisfaction from economizing measures. It is also possible that improvements in service could have an interaction that decreases satisfaction in some other aspect of service. This suggests that considerable caution should be taken before acting on model recommendations, to consider interactions that may have been overlooked.

Although it would add considerable complexity, there is no reason that such interactions among attributes could not be introduced to the model. The challenge would be to identify the interactions and to estimate their magnitude.

Linearity of the Preference Profiles

An assumption incorporated into the model presented here is that overall satisfaction is a linear sum of satisfaction with the individual attributes. This assumption is suspect from a psychological standpoint. It seems plausible that as satisfaction gets very high it becomes increasingly difficult to bring about improvements. At the low end of satisfaction the same resistance to improvement may exist, creating a tilted "S"-shaped function, or satisfaction may improve at a decreasing rate throughout its range, creating a downward-angled and stretched-out "C"-shaped function. It is even possible that the function is step-shaped, with alternating steep and shallow portions. Without clear evidence of an alternative functional form we used a linear function. It is easiest to manipulate and serves as a reasonably good approximation of the true function in light of the other limitations of the model. Other functions could also be modeled, however, either by transforming variables or by using a nonlinear programming approach.

SENSITIVITY TESTING

An example of sensitivity testing was given in Chapter 3, where we altered the estimated effect of service changes to see whether similar recommendations would be produced by the model. Sensitivity testing could also be performed with respect to the costs of service changes, where there is inherent uncertainty. We consider such testing an essential aspect of customer satisfaction research.

Ideally utility decision makers would attach a level of uncertainty to each of their estimates, cost and impact, and then each estimate could be tested at its high and low limit. This would entail an unwieldy if not impossible burden of running and rerunning the analysis, systematically altering each of the estimates and keeping track of changes in the recommended attribute changes. A streamlined approach, akin to what we performed for the second "pessimistic" scenario in the Chapter 3 test, would be to target only the more critical attributes for sensitivity testing.

In the end—and this is a point that cannot be stressed strongly enough—the model is not a substitute for good judgment on the part of utility management. Sensitivity testing can offer reassurance that a signal from the model to increase or decrease service is robust with respect to underlying assumptions, but it cannot substitute for the need ultimately to affirm that the change makes sense.

PARTIAL IMPLEMENTATION OF THE METHOD

Although the method developed here provides powerful insight into the relationship between customer satisfaction and costs, it may not be practical for every utility to implement all phases of research. Each phase provides a particular kind of information that should be of considerable value to managers on its own—though of course the information is most useful when it is combined. In this section we discuss options for partial implementation of the method.

Focus Groups Only

It would not be misguided for a utility to rely solely on focus groups for input from their customers. This method gets closest to customers' actual experience of service. Its main limitations are only uncertainty about generalization from the findings, and the lack of a quantitative purchase on the subject.

If a utility were to use focus groups as their main source of information we would recommend an effort be made to form a number of different groups based on specific experiences they had: some comprising people who were known to have had field service, others with people known to have phoned in to a customer service representative, yet others who had walked in to the offices, etc.—as many different experience-based groups (two groups for each experience) as the available budget would allow. It would be important not to reveal ahead of time the reason participants had been chosen, nor even that it was the water utility sponsoring the research. As in the focus groups described in Chapter 2, the discussion would only gradually be directed toward the specific experiences customers had with the water utility. A well-designed study could gather information about overall image of the utility that would be useful for public relations efforts, as well as more specific information about expectations for good service.

Survey Only

Another approach to partial implementation of the method would be to rely on the findings of the current research for qualitative insights and implement only the customer satisfaction survey. This would give decision makers insight into current satisfaction with specific aspects of service, and also estimate the rate of satisfaction change for increases and decreases in service, a useful indicator in itself of where efforts should be directed. We would expect utilities to add and delete service attributes, depending on which issues they perceive to be particularly salient to their operations. Some possible additions were suggested in the discussions of service categories in Chapter 3.

We suggested earlier that the response rate for the survey could be improved somewhat if two forms were developed, dividing attributes into two sets and randomly administering the different forms to the sample of customers. Another approach that would increase the efficiency of the survey, if the needed information were available, would be to target some surveys to a random subset of customers who had experienced certain aspects of customer service, such as field service, walk-in service, etc. This might most effectively be used to get responses for relatively rare experiences that were of interest, such as the resolution of bill problems. Supposing that 5% of customers have a bill problem in a given year, a simple random sample of 600 would be expected to net only 30 such customers. If these customers could be identified prior to administration of the survey a larger subsample (say, 100) could be targeted to improve the precision of the findings.

Simplified Survey

Since the satisfaction profiles were so consistent across the five samples we surveyed, a utility might decide to assume its customer base had similar profiles and administer a simplified survey that did not ask respondents to rate different levels of service but only to identify the current level they receive. This would provide the same information as the full survey except there would be a degree of uncertainty about the rate at which service changes would affect satisfaction.

If utility managers were interested in making decisions based on point estimates from their own customers and the satisfaction profiles reported in the current report, it would be crucial that questions were worded exactly as they are in the survey instrument we have developed. We do not know how sensitive the profiles are to question wording, but the general experience in survey research is that subtle wording differences can have an unexpectedly dramatic affect on findings.

Cost Exercise Only

Even if none of the other parts of the method were undertaken, there would be value in conducting the cost exercise. This is not customer satisfaction research, of course—at least not research from the customer's perspective. By itself it cannot cut through the unacknowledged biases of perspective that all organizations are prone to. Nevertheless, we see a great deal of value in opening customer service to a systematic internal analysis and asking what measures might be taken to improve satisfaction, and what those measures might cost. If done collectively we see it as an opportunity to mine the organization for ideas about improved service. We have no doubt that many utilities already do this to some degree. We think it is also important to estimate the impact on customer satisfaction from each measure, no matter how uncertain the estimate, as a way of sorting measures on the basis of importance to overall satisfaction.

Our experience suggests it will be easier to elicit ideas for service improvements than for economizing measures. Without customer satisfaction research to support the decision, however, the risk of reducing service is substantially elevated and may not be feasible. This exposes a problem with the *cost-exercise only* approach. It provides no way to know what impact service changes are having on satisfaction. In effect, this observation brings us back to the key theme of this report, which is the importance of customer-oriented research. Without it, all changes in service are done only on the basis of assumptions about customer preferences.

CHAPTER 5
RECOMMENDATIONS TO UTILITIES

ALLOCATE NECESSARY RESOURCES

Water utilities should find the resources necessary to conduct customer satisfaction research.

In Chapter 1, we argued that industries only begin to get serious about developing insight into customer perspectives when they are compelled by circumstances. We named three such circumstances in the drinking water industry that operate simultaneously: improved customer service in other industries, an increasingly competitive environment, and an internal desire to excel in one's life work. Taken together, these point to the need for research.

We are aware that many utility managers believe they already have a good understanding of how their organization is perceived by customers, and we have no doubt that this understanding often is, in many respects, valid. When it comes from an unsystematic source, however, this kind of insight is prone to distortions and omissions. Well-designed customer satisfaction research is intended to break through the biases of perspective that all organizations have—most notably at the top—so that managers can more rationally and more completely confront the reality of their customer relations.

DEVELOP DETAILED OPERATIONAL QUESTIONS

Utilities should develop customer satisfaction surveys and focus groups that ask detailed questions about operations, and not the kind that ask only general questions about satisfaction with the utility. The latter type of research does not provide information that can readily be acted on. Indeed, we suspect there is a degree of skepticism about the value of customer satisfaction research that is a product of managers coming into contact with such general satisfaction surveys. If the current research has demonstrated nothing else, perhaps it has shown that surveys delving no deeper than "How satisfied are you with the customer service you receive?" are not state of the art.

DEVELOP A COMPREHENSIVE SURVEY

Develop customer satisfaction surveys that go beyond point-estimates of satisfaction. Again, this is a matter of producing information that provides direction for decisions. Surveys should reveal not only how well a utility currently is performing but also how well and how poorly it could otherwise be imagined to perform. This is the task of providing customers with a contrast to current performance, which we argued is essential to the very nature of satisfaction. That is, customers make the contrast when they experience satisfaction, so in some manner the contrast must be reproduced in the measurement of satisfaction. We have shown one approach to representing these contrasts; there could be others.

COLLECT APPROPRIATE COST DATA

Collect cost data that can be directly linked to customer satisfaction data. The goal is to estimate a marginal cost of customer satisfaction with respect to as many attributes of service as possible. Then changes can be made where the costs of service improvement are lowest.

AVOID COMPLACENCY

Utility managers should not assume that because overall ratings of customer satisfaction for their utility are high no significant improvements could be made. We refer the reader back to the findings of Chapter 3. The two utilities that were estimated to get the greatest benefit from service changes were D.C. WASA and Kissimmee Water Services Department. It is noteworthy that while high benefits to D.C. WASA can be attributed to its comparatively low current performance, Kissimmee had the *highest* overall customer satisfaction rating, yet it was predicted to gain an 8% increase in overall customer satisfaction from the recommended changes in service. In contrast, Nashville and Phoenix, which had marginally lower overall customer satisfaction ratings than Kissimmee, were predicted to net only 1% or 2% improvement in satisfaction from the recommended changes. The lesson we take from this finding is that even organizations that are currently performing very well may be able to do considerably better. Without research, it would be easy to become complacent about a good overall rating.

INTERPRET RESULTS CAREFULLY

If the optimization modeling proposed in this report is undertaken, or some other method similar to it, do not take the recommendations of the model to be authoritative on their own. Rather, treat them as one important source of information that should be confirmed by other information, such as recommendations from internal sources, the experiences of managers at other utilities, and the published literature on customer satisfaction. We recommend interpreting the model's indications for one-versus-two levels of service change as strong-versus-weak signals about the optimal direction of change; we would not take them as gospel, especially not without corroborating evidence from other sources.

APPENDIX A
USING FOCUS GROUPS FOR CUSTOMER SATISFCTION RESEARCH

In this appendix, we offer advice to utilities that want to either conduct their own focus group research or hire a professional research company to conduct the groups. There are a number of useful guides available that could also be consulted (see for instance Krueger and Casey 2000; Morgan 1997). The comments presented here are tailored to the needs and circumstances of water utilities that we have encountered in this and other research.

For the case studies, we conducted two focus groups in the service territory of each partner utility. At each location, we hired a professional focus group research firm to recruit participants. In three of the four cities, we used the firm's facility to conduct the groups. In Kissimmee, Florida, no focus group facility was available, so we held the interviews in a conference room at the Hyatt Orlando Hotel. Focus group sessions were conducted in the evening, typically scheduled for 6:00 p.m. and 8:00 p.m. Between nine and thirteen people participated in each group. The participants were homeowners, selected to represent a cross-section of ages and incomes and a balance of men and women.

Although the research conducted for this project sheds considerable light on the question of customer expectations, we recommend that utilities invest in focus group research with their own customers, if at all possible, to supplement the findings reported here. At each location where we conducted interviews, there were findings particular to the site that were of significant interest to the utility partner. We have no doubt this would be true at almost any location. Moreover, we have frequently heard from utility personnel that observing the groups is enlightening, sometimes rather painful because of unflattering opinions and mistaken impressions that are aired, but always a bracing dose of reality.

There are four phases to focus group research: designing a question guide, recruiting participants, conducting the groups, and analyzing results. We will consider each phase in turn.

Designing the Question Guide

The question guide for a focus group is a relatively short list of questions on the topics to be covered. Along with the main questions are specific probes on topics that are of importance. These are asked only if participants have not raised them spontaneously. It is important to understand that the direction of conversation and the order in which topics are covered should not be entirely under the control of the focus group moderator. The research findings partly consist of the direction in which participants take the conversation as new topics emerge.

The question guide used for this research is included in an appendix (see Appendix B). This could be used as is, or modified to suit particular needs. A general rule for focus group questioning is to begin with very general issues to see what specific themes the participants themselves raise. In designing the opening questions, it is important to consider the information participants will have about the purpose of the interview. For the current research, we told participants only that "We want to talk to people like yourself about the service they receive from one of the utilities in your area." Part of the reason for keeping the sponsorship vague is to avoid biasing the recruitment process. The other reason, however, is that we want the initial conversation to be uncluttered with specific expectations about the topic, to provide maximum freedom to bring up topics participants themselves think are important.

For this project we initially began with the following question:

> As a customer, what kind of treatment do you expect from the companies you do business with? [PROBE: What defines good customer service? PROBE for bad service & examples thereof.]

It soon became clear that this question was rich enough by itself that we could have continued the discussion of it for the entire interview. After four groups, we narrowed the initial question somewhat in order to move more quickly into more specific issues. We asked:

> My client is a public utility. What kind of service do you get from the utilities in this city? [PROBE: How does it compare to customer service at retail establishments? Generally speaking, do you consider it to be better, or worse, or the same? PROBE for what makes it better or worse.]

Since participants do not know which utility is sponsoring the research, this provides an opportunity to gauge awareness of and attitudes toward the water utility relative to others.

Once we identified that our client was the water utility we asked another broad question, to see what impressions are at the front of people's minds. We asked:

> When you think about your own water utility, what thoughts and impressions, if any, come to your mind?

Again, the effort is to provide a blank canvas on which the group will inscribe its opinions, attitudes, experiences, etc. To be sure, it may turn out that very few thoughts and impressions come to the minds of participants. This, in itself, is a finding.

Only at this point did we begin to focus the question guide on more specific topics. For this research, we divided the questions among broad categories of service that we had already identified: water quality, water quality information, field service experiences, call-in and walk-in experiences, billing experiences, and utility-community relations. It is critical that as questions become more focused and specific they not lead participants toward the researchers' own views or expectations by incorporating assumptions about the actions, thoughts, or feelings of customers. The leading effect of questions is sometimes subtle. For instance the question "What information would you like to see on your water bill?" is leading if "Do you read your water bill?" or related questions have not been asked first.

It is good practice to break into the discussion every 20 to 25 minutes with a task that is an alternative to the general pattern of question and response. For this research, we used paper-and-pencil ratings of service attribute importance and written descriptions of different levels of service (see Appendix C). These were designed to ensure that we had information needed for the survey design. Discussion of handouts, in this case Consumer Confidence Reports, is another method we employed to keep concentration high and the conversation progressing.

Recruiting Participants

Participants for focus groups could be recruited from utility customer lists. Alternatively, if a professional focus group organization does the recruiting, it typically maintains a list of people who are interested in participating in focus groups that could be used. It is often less expensive to recruit off these lists because the acceptance rate is much higher than if recruiting from general lists. Proper screening is needed, however, to ensure participants are indeed customers of the utility. For instance, apartment dwellers that do not pay for water should be eliminated. If such lists are used, we recommend requiring the recruiters to call only people who have not participated in a focus group for at least six months in order to avoid participants who have become habituated to the experience.

For most customer satisfaction focus groups, the primary consideration is recruiting a cross-section of the customer base, including quotas for education, age, sex, and other variables that might have a systematic effect on attitudes. In some cases it might be advisable to form a group among, for instance, people who have contacted customer service in the past twelve months.

We have already discussed the issue of what information to provide to prospective recruits about the purpose of the focus groups sessions. Our strategy is to say as little as possible, while saying enough to convince potential recruits that the purpose is legitimate and worth participating in. A cash incentive of $50 to $75 typically is offered for participation.

Conducting the Interviews

This facet of the method is too complex to do it justice in a discussion such as this, so we refer readers to texts on the topic such as those cited above. We offer only a few points of advice, based on experiences we have had with water utilities.

Focus groups should not be conducted on water utility property. Likewise, utility personnel should not moderate the discussion. It is normatively more difficult to voice criticisms directly to one's host. A neutral location and moderator greatly increase the willingness of participants to speak freely. A good moderator cultivates his or her neutrality in a variety of ways. For utility representatives, focus group facilities almost always have an area where the interviews can be watched without intruding on the interactions. If a focus group facility is not used, the interview can be viewed on closed-circuit video. Focus groups are nearly always video recorded for subsequent viewing.

Utility representatives who observe the interviews are often quite frustrated at hearing outlandish views of their operations voiced by the participants. Viewers should not be tempted to correct any mistaken notions held by the focus group participants—certainly not before the very end of the interview. Keep in mind that the purpose of the sessions is information gathering not public communication. This rule is especially difficult to follow when participants ask the moderator questions about the utility's service—another good reason to hire someone from outside the company who can truthfully say he or she does not know the answer, and turn the question back to the group. If the moderator begins providing information, the assumption of expertise shifts away from the participants, and they may become reticent to speak.

Analysis of the Findings

Focus groups can be analyzed in different ways for different purposes. We have found that the most important analysis comes immediately following an evening's interviews, when several observers and the moderator discuss their impressions. These impressions should be written down as soon as possible—if not the same evening then the first thing the next morning: freshness of recall is vital and fades quickly—and then the notes should be checked against a second viewing of the interview. In many cases, utilities will find this form of analysis is sufficient in itself. One value of conducting interviews over several evenings is that observations from the first evening can be tested and further probed in subsequent interviews.

It is important to keep in mind that the distribution of attitudes in the focus groups is not expected to be representative of the customer base as a whole. The method is not intended to accurately measure the percentage of customers holding a particular view. The sample is simply too small and the selection process not sufficiently random to ensure representativeness in this regard. There may, however, be strong signals that some attitudes are nearly universal, or nearly absent.

APPENDIX B
FOCUS GROUP QUESTION GUIDE

I. Introduction (3-4 minutes)

A. Purpose: Research – no sales
B. Confidentiality
C. Audio/video taping: client observation
D. Encourage participation—no right or wrong answers
E. Questions?

II. Participant Introductions (10 minutes)

A. First name
B. Occupation
C. Household size and make-up
D. Name a local business that provides good customer service

III. General Customer Satisfaction (15 minutes)

As a customer, what kind of treatment do you expect from the companies you do business with? [PROBE: What defines good customer service? PROBE for bad service & examples thereof.]

IV. Utility Satisfaction (10 minutes)

My client is a public utility. What kind of service do you get from the utilities in this city? [PROBE: How does it compare to customer service at retail establishments? Generally speaking, do you consider it to be better, or worse, or the same? PROBE for what makes it better or worse.]

V. Water Utility Satisfaction (15 minutes)

My client is your water utility, which is collaborating with a national foundation that does research into drinking water issues.

If you know it, would you write the name of your water utility on the piece of paper in front of you?

I'm traveling to a number of locations around the country, talking to water utility customers, trying to get a good sense of the kinds of customer service that people are receiving and the kinds of service they would like to receive.

A. When you think about your own water utility, [name of water utility], what **thoughts and impressions**, if any, come to your mind? [PROBE: Do you generally think they are doing a good job, or a poor job? Do they give you the service you expect? (Probe at length for reasons why.) What do they do well? Where do they not perform so well?]
B. When you're sizing up the job they do, what do you consider some of the **most important factors** that determine whether they are doing a good job? [PROBE: What are the most important things for them to get right? What makes these things important?)]
C. What if your water utility wanted to go **above and beyond** what you expect from them? What do you wish they would do? [Probe: Are there any services you would like them to perform that they don't?]

VI. Specific Water Utility Satisfaction Issues (60 minutes)

A. Water Quality
   1. How do you feel about the quality of your drinking water? [PROBE: How would you describe its Taste and Odor? What about Color & Clarity? How would you describe

your water pressure? [PROBE for whether each attribute meets their expectations. PROBE for the frequency or problems]

2. Do you have any concerns about the quality of your water? [PROBE: Is it clean and safe?]
3. How many of you drink bottled water? Why/Why not?

B. Water Quality Information

1. What kind of information do you receive from the utility about your water? Is this information important to you? Is the information you receive easy to understand?
2. What would you like to know about your water? [ideally/minimally]
3. How often would you like to receive information? [ideally/minimally]
4. Did you see the annual water quality report the utility put out last year? Did you look at it? Do you remember any of what it said?

PASS OUT CCR. Take a minute to look at this water quality report.

5. Is it easy to understand? How does it make you feel about the quality of your water? [PROBE: Does it make you more concerned, less concerned, or not influence your concern?]
6. Is this a good way to receive water quality information? Is there a way that you think would be better? Would you be willing to log onto the web to receive water quality information? Has anyone here used the web to get information about water quality or other facts pertaining to drinking water?

C. On-Site Service [Explain term]

Have any of you had a service representative from the water utility come to your home to do work? Have you all had service people from some company come to your home?

1. What are the factors that make this either a good or a bad experience? [PROBE: What would you like to tell your water utility about how to do it right? What kinds of things bother you?]
2. What's a reasonable length of time to wait before the service person arrives at your house to make repairs? [PROBE to discover *ideal, acceptable,* and *minimal* levels of service.]
3. What kind of an appointment window would you like to have? [PROBE to discover *ideal, acceptable,* and *minimal* levels of service.]
4. During what hours should service be available? [PROBE to discover *ideal, acceptable,* and *minimal* levels of service.]
5. How do you want the service worker to behave toward you [Probe for specifics.]

DO "10 PER" TRADE-OFF EXERCISE: I'm going to pass out a sheet of paper that lists eight features of service you would receive when a service representative from the water utility comes to your house. [READ LIST] We want to know how important they are, compared to one another. You have 80 points to distribute among the eight areas of service. The more points you give an area, the more important it is to you. If they are equally important, you can give each area 10 points. If one area is more important than the others, give it more points and take some points away from the other areas. The total should be 80 points. Using this system, how would you score the importance of these eight areas of service?

Did you find this task easy to do? Did you have trouble deciding how many points to give each feature? [PROBE for point allocations and ask for rationale.]

D. Customer Service [Explain term]
Has anybody ever phoned his or her water utility with a question comment or complaint? [If yes: What was the purpose of your call? Was your call handled satisfactorily?]

1. What kind of treatment would you like when you phone a large company or a utility for service? What kinds of things bother you? [PROBE to discover *ideal, acceptable,* and *minimal* levels of service.]
2. A lot of utilities have a hard time deciding what are the best hours to keep their offices open. They want to please customers but it's expensive to offer service if customers don't use it. From your point of view, what hours would you like to see for customer service? Is this an important issue to you? [PROBE to discover *ideal, acceptable,* and *minimal* levels of service.]
3. More and more often, when you call a business on the telephone your call will be answered by an automated system. What are your thoughts about these automated phone answering systems? [PROBE to discover *ideal, acceptable,* and *minimal* levels of service.]

DO "10 PER" TRADE-OFF EXERCISE

E. Billing

1. What method do use to you pay your bill? [PROBE: by mail, in person, or some other way?] Is this convenient for you or would you prefer some other system?
2. Do you read your water bill? What kind of information does it contain? What information would you like to see on your water bill? [PROBE: Stuffers]
PASS OUT BILL FACSIMILE
3. I'm passing around a facsimile of the water bill you receive. Is it easy to read? Easy to understand? [Probe for why/why not]
4. Do you expect your bill to be accurate? Is your bill accurate? [PROBE to discover *ideal, acceptable,* and *minimal* levels of service.]
5. How often should the bill come? Would you mind if your bill came only every 3 or 6 months?
6. Should your water utility read your meter every bill cycle or would it be acceptable to estimate your bill on some months? [PROBE: Would you have any concerns about estimated bills?]

DO "10 PER" TRADE-OFF EXERCISE

F. Other Services

1. Are there other services you would like to see your water utility provide? [PROBE: Conservation assistance; information; bottled water; home water treatment]

G. Corporate Citizenship

1. Do you expect your water utility to play any broader role in your community in addition to providing water and wastewater services? What role?/Why not? [PROBE: Are there any kinds of outreach programs you would like to see? Education? Environmental? Do you consider those an important part of your water utility's job?]
2. What do you think your water utility should do to meet the future water needs of your city?

VII. Wrap-Up/Final Comments

APPENDIX C
SERVICE TRADE-OFF EXERCISES FOR FOCUS GROUPS

First Name________________________________

Billing Trade-Off Exercise

Think about the following five areas of service related to billing. We want to know how important they are, compared to one another. You have **50 points to distribute among the five areas of service.** The more points you give an area, the more important it is to you. If they are equally important, you can give each area 10 points. If one area is more important than the others, give it more points and take some points away from the other areas. The total should be 50 points. Using this system, how would you score the importance of these five areas of service?

_____ How easy it is to understand your bill
_____ Number of errors on your bill
_____ Resolution of problems you have with your bill
_____ How often you are billed
_____ Range of options for paying your bill (credit card, internet, etc.)
= 50 points total

First Name______________________________

On-Site Service Trade-Off Exercise

Think about the following eight areas of service related to on-site service. We want to know how important they are, compared to one another. You have **80 points to distribute among the eight areas of service.** The more points you give an area, the more important it is to you. If they are equally important, you can give each area 10 points. If one area is more important than the others, give it more points and take some points away from the other areas. The total should be 80 points. Using this system, how would you score the importance of these eight areas of service?

_____ Hours that on-site service is available
_____ Restoration of your property following on-site service
_____ Length of the appointment window
_____ Response time for on-site service
_____ On-schedule arrival of service representative
_____ Courtesy of service representative
_____ Knowledge of service representative
_____ Follow-up checks after on-site service
= 80 points total

First Name______________________________

Customer Service Trade-Off Exercise

Think about the following five areas of service related to customer service. We want to know how important they are, compared to one another. You have **50 points to distribute among the five areas of service.** The more points you give an area, the more important it is to you. If they are equally important, you can give each area 10 points. If one area is more important than the others, give it more points and take some points away from the other areas. The total should be 50 points. Using this system, how would you score the importance of these five areas of service?

_____ Hours that customer service is available
_____ Easy access to a live person when phoning
_____ Courtesy of service representative
_____ Knowledge of service representative
_____ On-line customer service available
=50 points total

APPENDIX D
CUSTOMER SATISFACTION SURVEY INSTRUMENT

START HERE

Do you get your water from a private well? *(Circle one number)*

1 Yes ➡ This survey only applies to persons who get their water from a public water supply. Please circle yes and return the survey in the postage-paid envelope. Thank you.

2 No ➡ Continue to question 1.

•

This survey should be filled out by the adult (18 years of age or older) most knowledgeable about your households' water use.

General Satisfaction

1 **Overall, how satisfied are you with your water utility? On a 100% scale, where 0% means not at all satisfied and 100% means completely satisfied, rate your overall satisfaction with your water utility.** *(Circle one number)*

Not at all satisfied										Completely satisfied
0%	10%	20%	30%	40%	50%	60%	70%	80%	90%	100%

2 **Approximately how much is your typical bill *for water service only*?** *(Do not include sewer charges, garbage collection, or other charges that are on your water bill.)*

$ ________ Typical charge per payment period for water service

3 **How satisfied are you with the price you are currently paying for *water service*?** *(Circle one number)*

Not at all satisfied										Completely satisfied
0%	10%	20%	30%	40%	50%	60%	70%	80%	90%	100%

Water Quality and Reliability

4 **In the past 12 months, have you experienced any of the following at your home?** *(Circle one number for each)*

Low water pressure	1 Yes	2 No
Unpleasant tasting tap water	1 Yes	2 No
Tap water odor or smell	1 Yes	2 No
Discolored or cloudy tap water	1 Yes	2 No
Concern over the healthfulness of your tap water	1 Yes	2 No

5 **How satisfied are you with each of the following?** *(Circle one number for each)*

		Not at all satisfied										Completely satisfied
a.	the water pressure at your tap	0	10	20	30	40	50	60	70	80	90	100%
b.	the taste of your tap water	0	10	20	30	40	50	60	70	80	90	100%
c.	the odor or smell of your tap water	0	10	20	30	40	50	60	70	80	90	100%
d.	the color and clarity of your tap water	0	10	20	30	40	50	60	70	80	90	100%
e.	the healthfulness of your tap water	0	10	20	30	40	50	60	70	80	90	100%

6 **Which of the following best describes your water drinking habits at home?** *(Circle one number)*

1 I only drink tap water (SKIP TO Question 8)
2 I drink tap water more often than bottled water
3 I drink tap water and bottled water equally
4 I drink bottled water more often than tap water
5 I only drink bottled water

7 **What are the main reasons you drink bottled water at home?** *(Circle all that apply)*

1 Convenience
2 Healthfulness
3 Taste
4 Other (please specify ________________)

8 **Do you use a home water filter of any kind for your tap water?** *(Circle one number)*

1 Yes → Which of the following best describes how often you use a home water filter? *(Circle one number)*
  1 Always
  2 Sometimes
  3 Rarely
2 No
3 Don't know

Billing Experiences

9 **About how many times in the past 12 months has each of the following happened?** *(Write a number for each experience; if none, write 0)*

Number of times in past 12 months

______You called your water utility to ask a question about your water bill.
______You had a problem with your water bill.

If you have had any type of problem with your bill in the past 12 months, please describe the problem.

____

____

10 **How satisfied are you with how easy it is to understand your water bill?** *(Circle one number)*

Not at all satisfied										Completely satisfied
0%	10%	20%	30%	40%	50%	60%	70%	80%	90%	100%

The following questions describe different service experiences that a water utility customer might have. Please rate ***how satisfied you would be if you had these experiences*** by circling a number on the scale next to **each item**. Then tell us which item most closely reflects your own experience by ***circling the appropriate letter***.

11 **If you were to find a problem on your water bill, here are three ways the problem might be resolved. How satisfied would you be in each situation?** *(Circle one number for each)*

		Not at all satisfied										Completely satisfied
a.	The billing problem is resolved after one contact with the utility and with no negotiations	0	10	20	30	40	50	60	70	80	90	100%
b.	The billing problem is resolved after two or three contacts with the utility and minor negotiations.......	0	10	20	30	40	50	60	70	80	90	100%
c.	The billing problem is resolved after more than 3 contacts with the utility and difficult negotiations ...	0	10	20	30	40	50	60	70	80	90	100%

Which of the above resolutions best represents the way billing problems have been resolved with your utility? If you have never had a billing problem, check this box □ and circle the letter that best describes how you think billing problems would be resolved with your utility. *(Circle one letter in the box below – these letters represent the 3 situations above in question 11)*

a. b. c.

12 **Here are four different schedules on which your water bill might be sent. How satisfied would you be with each of the following?** *(Circle one number for each)*

		Not at all satisfied										Completely satisfied
a.	The water bill is sent every month........................	0	10	20	30	40	50	60	70	80	90	100%
b.	The water bill is sent every two months	0	10	20	30	40	50	60	70	80	90	100%
c.	The water bill is sent every three months	0	10	20	30	40	50	60	70	80	90	100%
d.	The water bill is sent every six months................	0	10	20	30	40	50	60	70	80	90	100%

Which of the above schedules best reflects the *actual* schedule on which your bill is sent to you? If you do not know how often your water bill is sent, check this box □ and circle the letter that best describes how often you think your water bill is sent. *(Circle one letter in the box below)*

a. b. c. d.

13 **Thinking about the aspects of water billing services discussed above (billing frequency and billing problems) and anything else about billing you think is important, overall how satisfied are you with your actual water billing experiences?** *(Circle one number)*

Not at all Satisfied										Completely Satisfied
0%	10%	20%	30%	40%	50%	60%	70%	80%	90%	100%

Field Service Experiences

14 **The following list represents non-emergency field service that your water utility provides. How many times in the past 12 months has someone from your water utility come to your home or property for each of the following reasons?** *(Write in the number of visits for each reason; if none, write 0)*

Number of times in past 12 months

_____Install or repair/maintain water meter (do <u>not</u> include routine meter readings)
_____Investigate and/or repair a leak in the water line
_____Restore service following a water cut-off for late payment of your bill
_____To obtain a meter reading to start or stop water service
_____Other reasons (please specify ______________________________)

The following describes different experiences you could have when someone from the water utility comes to your home to install or repair equipment. Please rate ***how satisfied you <u>would be</u> if you had these experiences*** by circling a number on the scale next to each item. Then tell us which most closely reflects your own experience by ***circling the appropriate letter***.

15 **Here are four different hours of operation a water utility might be open for non-emergency field service. How satisfied would you be with each schedule of hours?** *(Circle one number for each)*

	Not at all satisfied										Completely satisfied
a. Field service (non-emergency) is available 24 hours/day, 7 days/week..	0	10	20	30	40	50	60	70	80	90	100%
b. Field service (non-emergency) is available 12 hours/day (for example, 8 AM to 8 PM), 7 days/week..	0	10	20	30	40	50	60	70	80	90	100%
c. Field service (non-emergency) is available 8 hours/day (for example, 9 AM to 5 PM), Monday - Friday	0	10	20	30	40	50	60	70	80	90	100%
d. Field service (non-emergency) is available 4 hours/day (for example, 7 AM to 11 AM), Monday - Friday..	0	10	20	30	40	50	60	70	80	90	100%

⬆ **Which schedule best represents your water utility's *actual* hours of operation for non-emergency field service available to you? If you have <u>never</u> had field service, check this box □ and circle the letter that best describes what hours you think would be available.** *(Circle the letter in the box below)*

a. b. c. d.

16 **Here are three different appointment windows a water utility might offer when it schedules field service at your property for non-emergencies. How satisfied would you be with each of the following?** *(Circle one number for each)*

	Not at all satisfied										Completely satisfied
a. The appointment is scheduled within a 2 hour window	0	10	20	30	40	50	60	70	80	90	100%
b. The appointment is scheduled within a 4 hour window	0	10	20	30	40	50	60	70	80	90	100%
c. The day of the appointment is scheduled, but not a time during the day	0	10	20	30	40	50	60	70	80	90	100%

Which of the above best represents the *actual* appointment window your water utility offers? If you have never had field service at your residence, check this box □ and circle the letter that best describes the appointment window you think you would be offered by your water utility. *(Circle the letter in the box below)*

a. b. c.

17 **Here are four different time periods that might elapse between the time you contact your water utility and the time a field service representative comes to your home to make repairs for non-emergencies. How satisfied would you be with each of the following?** *(Circle one number for each)*

	Not at all satisfied										Completely satisfied
a. The service representative arrives within 4 hours after your call for service	0	10	20	30	40	50	60	70	80	90	100%
b. The service representative arrives on the same day as your call for service	0	10	20	30	40	50	60	70	80	90	100%
c. The service representative arrives the next day after your call for service	0	10	20	30	40	50	60	70	80	90	100%
d. The service representative arrives within 3 days after your call for service	0	10	20	30	40	50	60	70	80	90	100%

Which of the above best represents the *actual* response time you receive from your water utility? If you have never had non-emergency field service, check this box □ and circle the letter that best describes how you think your utility would respond. *(Circle the letter in the box below)*

a. b. c. d.

18 **Here are four different forms of arrival behavior you might experience from a water utility when a service representative comes to your property to make repairs for non-emergencies. How satisfied would you be with each of the following?** *(Circle one number for each)*

	Not at all satisfied										Completely satisfied
a. The service representative arrives on time and phones beforehand to let you know they are coming	0	10	20	30	40	50	60	70	80	90	100%
b. The service representative arrives on time without phoning beforehand	0	10	20	30	40	50	60	70	80	90	100%
c. The service representative arrives late but phones beforehand to let you know	0	10	20	30	40	50	60	70	80	90	100%
d. The service representative arrives late without phoning to let you know	0	10	20	30	40	50	60	70	80	90	100%

Which of the above best represents the *actual* arrival behavior you have experienced? If you have never had field service, check this box ☐ and circle the letter that best describes how you think the arrival behavior would be. *(Circle the letter in the box below)*

a. b. c. d.

19 Here are four different levels of professionalism you might experience from water utility service representatives when they come to your property to make repairs for non-emergencies. How satisfied would you be with each of the following? *(Circle one number for each)*

	Not at all satisfied										Completely satisfied
a. The service representatives give clear information about what they are doing, listen carefully to what you say, and are polite and friendly	0	10	20	30	40	50	60	70	80	90	100%
b. The service representatives do not give clear information about what they are doing, but listen carefully to what you say, and are polite and friendly	0	10	20	30	40	50	60	70	80	90	100%
c. The service representatives do not give clear information about what they are doing, do not listen carefully to what you say, but are polite and friendly	0	10	20	30	40	50	60	70	80	90	100%
d. The service representatives do not give clear information about what they are doing, do not listen carefully to what you say, and are a little rude or unfriendly	0	10	20	30	40	50	60	70	80	90	100%

Which of the above best represents the *actual* level of professionalism you have experienced? If you have never had field service, check this box ☐ and circle the letter that best describes the level of professionalism you think you would receive. *(Circle the letter in the box below)*

a. b. c. d.

20 **Here are four different ways water utility service representatives might follow up after they come to your property to make repairs for non-emergencies. How satisfied would you be with each of the following?** *(Circle one number for each)*

	Not at all satisfied										Completely satisfied
a. The service representative follows up in writing and with a phone call	0	10	20	30	40	50	60	70	80	90	100%
b. The service representative follows up with a phone call but not in writing	0	10	20	30	40	50	60	70	80	90	100%
c. The service representative follows up in writing but not with a phone call	0	10	20	30	40	50	60	70	80	90	100%
d. The service representative does not follow up	0	10	20	30	40	50	60	70	80	90	100%

Which of the above best represents the *actual* way the service representative has followed up? If you have never had field service, check this box □ and circle the letter that best describes the way you think the service representative would follow up. *(Circle the letter in the box below)*

a. b. c. d.

21 **Thinking about all of the aspects of field service discussed above (hours of operation, restoring property, appointment window, arrival times, arrival behavior, professionalism, and follow up), overall, how satisfied are you with your actual field service experiences with your water utility?** *(Circle one number)*

Not at all satisfied										Completely satisfied
0%	10%	20%	30%	40%	50%	60%	70%	80%	90%	100%

Customer Service Experiences

22 **How many times in past 12 months have you contacted your water utility using the following methods?** *(Write in number of times; if none, write 0)*

Number of times in past 12 months

_____Phoned your water utility
_____Visited your water utility office(s)
_____Logged onto your water utility's web site

The following describe different types of experiences with customer service that you might have when you call or walk in to the office of your water utility (for example, to get information about your account, to pay a bill, etc.). Please rate ***how satisfied you would be if you had these experiences*** by circling a number on the scale next to each item. Then tell us which most closely reflects your own experience by ***circling the appropriate letter***.

23 **Here are four different customer service hours a water utility might offer to people who want to speak with a customer service representative either in person or by phone. How satisfied would you be with each of the following?** *(Circle one number for each)*

	Not at all satisfied										Completely satisfied
a. Customer service is available 24 hours/day, 7 days/week	0	10	20	30	40	50	60	70	80	90	100%
b. Customer service is available 12 hours/day (for example, 7 AM to 7 PM), Monday - Friday, and some weekend hours	0	10	20	30	40	50	60	70	80	90	100%
c. Customer service is available 12 hours/day (for example, 7 AM to 7 PM), Monday - Friday	0	10	20	30	40	50	60	70	80	90	100%
d. Customer service is available 8 hours/day (for example, 9 AM to 5 PM), Monday - Friday	0	10	20	30	40	50	60	70	80	90	100%

Which of the above best represents the *actual* hours of customer service available to you? If you do not know the hours of customer service, check this box ☐ and circle the letter that best describes the hours of customers service you think are offered. *(Circle the letter in the box below)*

a. b. c. d.

24 **Here are three different levels of authority a customer service representative might have about a request for information or action when you call or walk in to your water utility office. How satisfied would you be with each of the following?** *(Circle one number for each)*

	Not at all satisfied										Completely satisfied
a. The customer service representative is authorized to make decisions and completes the job easily	0	10	20	30	40	50	60	70	80	90	100%
b. The customer service representative has to seek authorization from others but completes the job easily	0	10	20	30	40	50	60	70	80	90	100%
c. The customer service representative has to seek authorization from others and completes the job with difficulty	0	10	20	30	40	50	60	70	80	90	100%

Which of the above best represents the *actual* level of authority you have experienced? If you have <u>never</u> called or walked into customer service check this box ☐ and circle the letter that best describes the level of authority you think you would receive. *(Circle the letter in the box below)*

a. b. c.

25 **Here are four different levels of professionalism you might experience from a customer service representative when you call or walk into the office. How satisfied would you be with each of the following?** *(Circle one number for each)*

	Not at all satisfied										Completely satisfied
a. The customer service representative is polite and friendly, fully explains utility policy and listens carefully to what you say.	0	10	20	30	40	50	60	70	80	90	100%
b. The customer service representative is polite and friendly, does not explain utility policy but listens carefully to what you say	0	10	20	30	40	50	60	70	80	90	100%
c. The customer service representative is polite and friendly, does not explain utility policy and does not listen carefully to what you say.	0	10	20	30	40	50	60	70	80	90	100%
d. The customer service representative is a little rude or unfriendly, does not explain utility policy and does not listen carefully to what you say.	0	10	20	30	40	50	60	70	80	90	100%

Which of the above best represents the *actual* level of professionalism you have experienced? If you have never called or walked into the customer service office of your water utility, check this box ☐ and circle the letter that best describes the level of professionalism you think you would receive. *(Circle the letter in the box below)*

a. b. c. d.

26 **Here are three experiences you might have on the telephone when you call a customer service representative at your water utility. How satisfied would you be with each of the following?** *(Circle one number for each)*

	Not at all satisfied										Completely satisfied
a. An automated system answers the phone. You are able to reach a customer service representative with one input from a short list of options.	0	10	20	30	40	50	60	70	80	90	100%
b. An automated system answers the phone. You are able to reach a customer service representative after three inputs from a long lists of options.	0	10	20	30	40	50	60	70	80	90	100%
c. An automated system answers the phone. You cannot reach a customer service representative.	0	10	20	30	40	50	60	70	80	90	100%

Which of the above best represents the *actual* experience you have had when calling customer service at your water utility? If you have never called customer service, check this box ☐ and circle the letter that best describes what you think your experience would be. *(Circle the letter in the box below)*

a. b. c.

27 **Here are four different lengths of time you might have to wait on hold before speaking to a customer service representative when you telephone the water utility office. How satisfied would you be with each of the following?** *(Circle one number for each)*

	Not at all satisfied										Completely satisfied
a. You do not have to wait on hold. You speak to a customer service representative right away.	0	10	20	30	40	50	60	70	80	90	100%
b. You are on hold for 20 seconds before speaking to a customer service representative	0	10	20	30	40	50	60	70	80	90	100%
c. You are on hold for 45 seconds before speaking to a customer service representative	0	10	20	30	40	50	60	70	80	90	100%
d. You are on hold for 120 seconds (2 minutes) before speaking to a customer service representative...	0	10	20	30	40	50	60	70	80	90	100%

Which of the above best represents the *actual* hold time you have experienced? If you have <u>never</u> telephoned customer service at your water utility, check this box □ and circle the letter that best describes what you think the hold time would be. *(Circle one letter in the box below)*

a. b. c. d.

28 **Here are four different lengths of time you might have to wait in line before being served by a customer service representative when you walk in to the water utility office. How satisfied would you be with each of the following?** *(Circle one number for each)*

	Not at all satisfied										Completely satisfied
a. You do not have to wait in line. A customer service representative serves you right away........	0	10	20	30	40	50	60	70	80	90	100%
b. You have to wait in line for 1 minute before a customer service representative serves you.	0	10	20	30	40	50	60	70	80	90	100%
c. You have to wait in line for 5 minutes before a customer service representative serves you.	0	10	20	30	40	50	60	70	80	90	100%
d. You have to wait in line for 10 minutes before a customer service representative serves you.	0	10	20	30	40	50	60	70	80	90	100%

Which of the above best represents the *actual* time you have had to wait in line? If you have <u>never</u> visited the service office at your utility, check this box □ and circle the letter that best describes the wait time you think you would receive. *(Circle one letter in the box below)*

a. b. c. d.

29 **Thinking about all of the aspects of customer service discussed above (customer service hours, knowledge, professionalism, automated phone system, wait in line and wait on hold), overall, how satisfied are you with your actual customer service experiences?** *(Circle one number)*

Not at all satisfied										Completely satisfied
0%	10%	20%	30%	40%	50%	60%	70%	80%	90%	100%

Utility Communications

30 **In the past 12 months, which of the following do you recall?** *(Circle all letters that apply)*

a. Reading an informational bill insert that comes with your water bill
b. Reading a water quality report produced by your water utility
c. Seeing on TV a public service announcement sponsored by your water utility
d. Hearing on the radio a public service announcement sponsored by your water utility
e. Seeing a billboard sponsored by your water utility
f. Seeing or hearing a news story about water quality issues related to your own water utility
g. Seeing or hearing a news story about water quality issues related to another water utility, either in your state or nationally
h. Logging on to the internet to get information about your water utility
i. None of the above

31 **Below are a variety of different ways a water utility might communicate with the public about its operations and about issues of special interest to its customers. Please rank the two most important ways and the least important way you believe your utility should communicate with the public. Use a "1" to indicate the most important, a "2" to indicate second most important and an "X" to indicate the least important.**

Rank (1 is most important, 2 is second most important and X is least important)

____ billboards
____ bill inserts
____ internet
____ mailings separate from water bill
____ newspapers
____ radio
____ television

32 **What kind of outreach event would you prefer your water utility use to inform and educate the public about water issues?** *(Circle one letter)*

a. clean-water festivals
b. community presentations
c. events and research opportunities for students
d. open houses at the treatment plant
e. school lectures and demonstrations
f. other (please specify ____)
g. none of the above

The following describes different levels of ***utility communications***. Please rate ***how satisfied you would be*** with these by circling a number on the scale next to each item. Then tell us which most closely reflects your own experience by ***circling the appropriate letter***.

33 **Here are four descriptions of how often a water utility might communicate with the public about its operations and about issues of special interest to its customers. How satisfied would you be with each of the following?** *(Circle one number for each)*

	Not at all satisfied										Completely satisfied
a. The utility communicates monthly	0	10	20	30	40	50	60	70	80	90	100%
b. The utility communicates 4 times/year	0	10	20	30	40	50	60	70	80	90	100%
c. The utility communicates once/year	0	10	20	30	40	50	60	70	80	90	100%
d. The utility communicates only in emergencies	0	10	20	30	40	50	60	70	80	90	100%

Which of the above best represents the *actual* way your water utility communicates with the public? If you do not know how often your utility communicates, check this box □ and circle the letter that best describes how often you think the utility communicates. *(Circle the letter in the box below)*

a. b. c. d.

34 **Here are four schedules on which a water utility might distribute routine (non-emergency) information about water quality. How satisfied would you be with each of the following?** *(Circle one number for each)*

	Not at all satisfied										Completely satisfied
a. Water quality information is distributed in the mail monthly	0	10	20	30	40	50	60	70	80	90	100%
b. Water quality information is distributed in the mail every three months	0	10	20	30	40	50	60	70	80	90	100%
c. Water quality information is distributed in the mail once a year	0	10	20	30	40	50	60	70	80	90	100%
d. Water quality information is available on the internet	0	10	20	30	40	50	60	70	80	90	100%

Which of the above best represents the *actual* way your utility distributes routine water quality information? If you do not know how often your utility distributes routine information, check this box □ and circle the letter that best describes how often you think the information is distributed. *(Circle the letter in the box below)*

a. b. c. d.

35 **Here are four different levels of information from your water utility you might be able to get on the internet. How satisfied would you be with each of the following?** *(Circle one number for each)*

	Not at all satisfied										Completely satisfied
a. Individual account information, bill payment history and general information	0	10	20	30	40	50	60	70	80	90	100%
b. Individual account information and general information	0	10	20	30	40	50	60	70	80	90	100%
c. Only general information	0	10	20	30	40	50	60	70	80	90	100%
d. No internet information	0	10	20	30	40	50	60	70	80	90	100%

Which of the above best represents the *actual* level of information your utility distributes on the internet? If you do not know what your utility distributes, check this box □ and circle the letter that best describes what you think is distributed. *(Circle the letter in the box below)*

a. b. c. d.

36 **Considering all of the aspects of public communications discussed above, how satisfied are you with the way your water utility communicates with the public?** *(Circle one number)*

Not at all satisfied										Completely satisfied
0%	10%	20%	30%	40%	50%	60%	70%	80%	90%	100%

About Your Household

This final series of questions asks about your household's characteristics. Your responses will be kept confidential and combined with those of others.

37 **What type of home do you live in?** *(Circle one number)*

1 Single family home
2 Apartment or condominium (2 or more units)
3 Other (please specify ______________________)

38 **In what year was your home built?** *(Write in year)*

______ Year home was built

39 **Do you own or rent your home?** *(Circle one number)*

1 Own
2 Rent

40 **Including yourself, how many people live in your home year-round?** *(Write in numbers on the appropriate lines)*

Number

____ Under 5 years old
____ 5 to 17 years old
____ 18 to 24 years old
____ 25 to 34 years old
____ 35 to 44 years old
____ 45 to 54 years old
____ 55 to 64 years old
____ 65 or more years old

41 **Which of the following categories best describes your age?** *(Circle one number)*

1 Under 18
2 18 to 24 years old
3 25 to 34 years old
4 35 to 44 years old
5 45 to 54 years old
6 55 to 64 years old
7 65 or more years old

42 **What is the highest level of education you have completed?** (*Circle one number)*

1 Some high school or less
2 Completed high school (or GED)
3 Some college or technical school, no degree
4 Two year college degree or technical school degree
5 Four year college graduate
6 Post-graduate work or degree

43 **Which of the following best describes your total 2001 household income before taxes and other deductions?** *(Circle one number)*

1 Less than $15,000
2 $15,000 to $24,999
3 $25,000 to $49,999
4 $50,000 to $74,999
5 $75,000 to $99,999
6 $100,000 or more

Thank you for your help with this important study.
Please return this survey in the postage-paid envelope to:

PA Consulting Group
2711 Allen Boulevard, Suite 200
Middleton, WI 53562

APPENDIX E
COST EXERCISE WORKSHEETS

Cost Exercise Worksheets

These worksheets are designed to help you estimate the costs of increasing and decreasing service performance for seventeen different attributes of customer service.

- We have divided the utility into four categories of service: billing, field service, customer service (phone and walk-in), and pubic communications.
- Each category is represented by attributes of service. These are ways in which the utility comes into contact with its customers.
- Each attribute of service can be delivered at different levels. We've asked your customers how satisfied they would be receiving service at each of these levels. (Your actual level of service may not be perfectly reflected in any of the categories. We have tried to define levels above and below what most utilities offer.)

For each attribute of service we would like you to tell us what steps you would take to improve service and what steps you would take to economize on service, how much the change would cost or how much savings would be achieved, and an estimate of the impact the change would have on customers' satisfaction (i.e., "complete," "substantial," "moderate," "modest," or "small.")

Service Category 1
Billing

1 If you were to find a problem on your water bill, here are three ways the problem might be resolved. How satisfied would you be in each situation?

		% Customers reporting	Average Satisfaction (100 pt. scale)
1.	The billing problem is resolved after one contact with the utility and with no negotiations	83	94
2.	The billing problem is resolved after two or three contacts with the utility and minor negotiations	13	42
3.	The billing problem is resolved after more than 3 contacts with the utility and difficult negotiations	4	16

Reported satisfaction? 82%

Resolving Bill Problems

A. What's the first thing you would do to **increase** service performance in this area?

Change	Cost
1.	
2.	
3.	
Total Cost	

How large an effect would you expect this to have on customer satisfaction?

_____Complete (moves about 9/10 of customers to a higher level of satisfaction)

_____Substantial (moves about 2/3 of customers to a higher level of satisfaction)

_____Moderate (moves about 1/2 of customers to a higher level of satisfaction)

_____Modest (moves about 1/3 of customers to a higher level of satisfaction)

_____Small (moves about 1/10 of customers to a higher level of satisfaction)

B. What's the second thing you would do to **increase** service performance in this area?

Change	Cost
1.	
2.	
3.	
Total Cost	

How large an effect would you expect this to have on customer satisfaction?

_____Complete (moves about 9/10 of customers to a higher level of satisfaction)

_____Substantial (moves about 2/3 of customers to a higher level of satisfaction)

_____Moderate (moves about 1/2 of customers to a higher level of satisfaction)

_____Modest (moves about 1/3 of customers to a higher level of satisfaction)

_____Small (moves about 1/10 of customers to a higher level of satisfaction)

C. What's the first thing you would do to **economize** on service performance in this area?

Change	Savings
1.	
2.	
3.	
Total Savings	

How large an effect would you expect this to have on customer satisfaction?

_____Complete (moves about 9/10 of customers to a lower level of satisfaction)

_____Substantial (moves about 2/3 of customers to a lower level of satisfaction)

_____Moderate (moves about 1/2 of customers to a lower level of satisfaction)

_____Modest (moves about 1/3 of customers to a lower level of satisfaction)

_____Small (moves about 1/10 of customers to a lower level of satisfaction)

D. What's the second thing you would do to **economize** on service performance in this area?

Change	Savings
1.	
2.	
3.	
Total Savings	

How large an effect would you expect this to have on customer satisfaction?

_____Complete (moves about 9/10 of customers to a lower level of satisfaction)

_____Substantial (moves about 2/3 of customers to a lower level of satisfaction)

_____Moderate (moves about 1/2 of customers to a lower level of satisfaction)

_____Modest (moves about 1/3 of customers to a lower level of satisfaction)

_____Small (moves about 1/10 of customers to a lower level of satisfaction)

2 **Here are four different schedules on which your water bill might be sent. How satisfied would you be with each of the following?** *(Circle one number for each)*

	% Customers reporting	Satisfaction (100 pt. scale)
1. The water bill is sent every month	96	96
2. The water bill is sent every two months..............................	3	42
3. The water bill is sent every three months............................	0	23
4. The water bill is sent every six months	1	8

Reported satisfaction? 95%

Schedule of Meter Reading

A. What's the first thing you would do to **increase** service performance in this area?

Change	Cost
1.	
2.	
3.	
Total Cost	

How large an effect would you expect this to have on customer satisfaction?

_____Complete (moves about 9/10 of customers to a higher level of satisfaction)

_____Substantial (moves about 2/3 of customers to a higher level of satisfaction)

_____Moderate (moves about 1/2 of customers to a higher level of satisfaction)

_____Modest (moves about 1/3 of customers to a higher level of satisfaction)

_____Small (moves about 1/10 of customers to a higher level of satisfaction)

B. What's the second thing you would do to **increase** service performance in this area?

Change	Cost
1.	
2.	
3.	
Total Cost	

How large an effect would you expect this to have on customer satisfaction?

_____Complete (moves about 9/10 of customers to a higher level of satisfaction)

_____Substantial (moves about 2/3 of customers to a higher level of satisfaction)

_____Moderate (moves about 1/2 of customers to a higher level of satisfaction)

_____Modest (moves about 1/3 of customers to a higher level of satisfaction)

_____Small (moves about 1/10 of customers to a higher level of satisfaction)

C. What's the first thing you would do to **economize** on service performance in this area?

Change	**Cost**
1.	
2.	
3.	
Total Cost	

How large an effect would you expect this to have on customer satisfaction?

_____Complete (moves about 9/10 of customers to a lower level of satisfaction)

_____Substantial (moves about 2/3 of customers to a lower level of satisfaction)

_____Moderate (moves about 1/2 of customers to a lower level of satisfaction)

_____Modest (moves about 1/3 of customers to a lower level of satisfaction)

_____Small (moves about 1/10 of customers to a lower level of satisfaction)

D. What's the second thing you would do to **economize** on service performance in this area?

Change	**Cost**
1.	
2.	
3.	
Total Cost	

How large an effect would you expect this to have on customer satisfaction?

_____Complete (moves about 9/10 of customers to a lower level of satisfaction)

_____Substantial (moves about 2/3 of customers to a lower level of satisfaction)

_____Moderate (moves about 1/2 of customers to a lower level of satisfaction)

_____Modest (moves about 1/3 of customers to a lower level of satisfaction)

_____Small (moves about 1/10 of customers to a lower level of satisfaction)

Service Category 2
Field Service

1 **Here are four different hours of operation a water utility might be open for non-emergency field service. How satisfied would you be with each schedule of hours?**

		% Customers reporting	Satisfaction (100 pt. scale)
1.	Field service (non-emergency) is available 24 hours/day, 7 days/week	40	85
2.	Field service (non-emergency) is available 12 hours/day (for example, 8 AM to 8 PM), 7 days/week..	24	75
3.	Field service (non-emergency) is available 8 hours/day (for example, 9 AM to 5 PM), Monday - Friday	33	51
4.	Field service (non-emergency) is available 4 hours/day (for example, 7 AM to 11 AM), Monday - Friday..	2	18

Reported satisfaction? 78%

Hours of Service

A. What's the first thing you would do to **increase** service performance in this area?

Change	Cost
1.	
2.	
3.	
Total Cost	

How large an effect would you expect this to have on customer satisfaction?

_____Complete (moves about 9/10 of customers to a higher level of satisfaction)

_____Substantial (moves about 2/3 of customers to a higher level of satisfaction)

_____Moderate (moves about 1/2 of customers to a higher level of satisfaction)

_____Modest (moves about 1/3 of customers to a higher level of satisfaction)

_____Small (moves about 1/10 of customers to a higher level of satisfaction)

B. What's the second thing you would do to **increase** service performance in this area?

Change	Cost
1.	
2.	
3.	
Total Cost	

How large an effect would you expect this to have on customer satisfaction?

_____Complete (moves about 9/10 of customers to a higher level of satisfaction)

_____Substantial (moves about 2/3 of customers to a higher level of satisfaction)

_____Moderate (moves about 1/2 of customers to a higher level of satisfaction)

_____Modest (moves about 1/3 of customers to a higher level of satisfaction)

_____Small (moves about 1/10 of customers to a higher level of satisfaction)

C. What's the first thing you would do to **economize** on service performance in this area?

Change	Savings
1.	
2.	
3.	
Total Savings	

How large an effect would you expect this to have on customer satisfaction?

_____Complete (moves about 9/10 of customers to a lower level of satisfaction)

_____Substantial (moves about 2/3 of customers to a lower level of satisfaction)

_____Moderate (moves about 1/2 of customers to a lower level of satisfaction)

_____Modest (moves about 1/3 of customers to a lower level of satisfaction)

_____Small (moves about 1/10 of customers to a lower level of satisfaction)

D. What's the second thing you would do to **economize** on service performance in this area?

Change	Savings
1.	
2.	
3.	
Total Savings	

How large an effect would you expect this to have on customer satisfaction?

_____Complete (moves about 9/10 of customers to a lower level of satisfaction)

_____Substantial (moves about 2/3 of customers to a lower level of satisfaction)

_____Moderate (moves about 1/2 of customers to a lower level of satisfaction)

_____Modest (moves about 1/3 of customers to a lower level of satisfaction)

_____Small (moves about 1/10 of customers to a lower level of satisfaction)

2 **Here are three different appointment windows a water utility might offer when it schedules field service at your property for non-emergencies. How satisfied would you be with each of the following?** *(Circle one number for each)*

		% Customers reporting	Satisfaction (100 pt. scale)
1.	The appointment is scheduled within a 2 hour window	56	91
2.	The appointment is scheduled within a 4 hour window	36	60
3.	The day of the appointment is scheduled, but not a time during the day	8	15

Reported satisfaction? 85%

Appointment Window

A. What's the first thing you would do to **increase** service performance in this area?

Change	Cost
1.	
2.	
3.	
Total Cost	

How large an effect would you expect this to have on customer satisfaction?

_____Complete (moves about 9/10 of customers to a higher level of satisfaction)

_____Substantial (moves about 2/3 of customers to a higher level of satisfaction)

_____Moderate (moves about 1/2 of customers to a higher level of satisfaction)

_____Modest (moves about 1/3 of customers to a higher level of satisfaction)

_____Small (moves about 1/10 of customers to a higher level of satisfaction)

B. What's the second thing you would do to **increase** service performance in this area?

Change	Cost
1.	
2.	
3.	
Total Cost	

How large an effect would you expect this to have on customer satisfaction?

_____Complete (moves about 9/10 of customers to a higher level of satisfaction)

_____Substantial (moves about 2/3 of customers to a higher level of satisfaction)

_____Moderate (moves about 1/2 of customers to a higher level of satisfaction)

_____Modest (moves about 1/3 of customers to a higher level of satisfaction)

_____Small (moves about 1/10 of customers to a higher level of satisfaction)

C. What's the first thing you would do to **economize** on service performance in this area?

Change	**Savings**
1.	
2.	
3.	
Total Savings	

How large an effect would you expect this to have on customer satisfaction?

_____Complete (moves about 9/10 of customers to a lower level of satisfaction)

_____Substantial (moves about 2/3 of customers to a lower level of satisfaction)

_____Moderate (moves about 1/2 of customers to a lower level of satisfaction)

_____Modest (moves about 1/3 of customers to a lower level of satisfaction)

_____Small (moves about 1/10 of customers to a lower level of satisfaction)

D. What's the second thing you would do to **economize** on service performance in this area?

Change	**Savings**
1.	
2.	
3.	
Total Savings	

How large an effect would you expect this to have on customer satisfaction?

_____Complete (moves about 9/10 of customers to a lower level of satisfaction)

_____Substantial (moves about 2/3 of customers to a lower level of satisfaction)

_____Moderate (moves about 1/2 of customers to a lower level of satisfaction)

_____Modest (moves about 1/3 of customers to a lower level of satisfaction)

_____Small (moves about 1/10 of customers to a lower level of satisfaction)

3 **Here are four different time periods that might elapse between the time you contact your water utility and the time a field service representative comes to your home to make repairs for non-emergencies. How satisfied would you be with each of the following?** *(Circle one number for each)*

		% Customers reporting	Satisfaction (100 pt. scale)
1.	The service representative arrives within 4 hours after your call for service	48	89
2.	The service representative arrives on the same day as your call for service	15	73
3.	The service representative arrives the next day after your call for service	28	44
4.	The service representative arrives within 3 days after your call for service	10	13

Reported satisfaction? 77%

Response Time

A. What's the first thing you would do to **increase** service performance in this area?

Change	Cost
1.	
2.	
3.	
Total Cost	

How large an effect would you expect this to have on customer satisfaction?

_____Complete (moves about 9/10 of customers to a higher level of satisfaction)

_____Substantial (moves about 2/3 of customers to a higher level of satisfaction)

_____Moderate (moves about 1/2 of customers to a higher level of satisfaction)

_____Modest (moves about 1/3 of customers to a higher level of satisfaction)

_____Small (moves about 1/10 of customers to a higher level of satisfaction)

B. What's the second thing you would do to **increase** service performance in this area?

Change	Cost
1.	
2.	
3.	
Total Cost	

How large an effect would you expect this to have on customer satisfaction?

_____Complete (moves about 9/10 of customers to a higher level of satisfaction)

_____Substantial (moves about 2/3 of customers to a higher level of satisfaction)

______Moderate (moves about 1/2 of customers to a higher level of satisfaction)

______Modest (moves about 1/3 of customers to a higher level of satisfaction)

______Small (moves about 1/10 of customers to a higher level of satisfaction)

C. What's the first thing you would do to **economize** on service performance in this area?

Change	**Savings**
1.	
2.	
3.	
Total Savings	

How large an effect would you expect this to have on customer satisfaction?

______Complete (moves about 9/10 of customers to a lower level of satisfaction)

______Substantial (moves about 2/3 of customers to a lower level of satisfaction)

______Moderate (moves about 1/2 of customers to a lower level of satisfaction)

______Modest (moves about 1/3 of customers to a lower level of satisfaction)

______Small (moves about 1/10 of customers to a lower level of satisfaction)

D. What's the second thing you would do to **economize** on service performance in this area?

Change	**Savings**
1.	
2.	
3.	
Total Savings	

How large an effect would you expect this to have on customer satisfaction?

______Complete (moves about 9/10 of customers to a lower level of satisfaction)

______Substantial (moves about 2/3 of customers to a lower level of satisfaction)

______Moderate (moves about 1/2 of customers to a lower level of satisfaction)

______Modest (moves about 1/3 of customers to a lower level of satisfaction)

______Small (moves about 1/10 of customers to a lower level of satisfaction)

4 **Here are four different forms of arrival behavior you might experience from a water utility when a service representative comes to your property to make repairs for non-emergencies. How satisfied would you be with each of the following?** *(Circle one number for each)*

		% Customers reporting	Satisfaction (100 pt. scale)
1.	The service representative arrives on time and phones beforehand to let you know they are coming	48	98
2.	The service representative arrives on time without phoning beforehand	45	79
3.	The service representative arrives late but phones beforehand to let you know..	6	74
4.	The service representative arrives late without phoning to let you know	0	16

Reported satisfaction? 95%

Arrival Etiquette

A. What's the first thing you would do to **increase** service performance in this area?

Change	Cost
1.	
2.	
3.	
Total Cost	

How large an effect would you expect this to have on customer satisfaction?

_____Complete (moves about 9/10 of customers to a higher level of satisfaction)

_____Substantial (moves about 2/3 of customers to a higher level of satisfaction)

_____Moderate (moves about 1/2 of customers to a higher level of satisfaction)

_____Modest (moves about 1/3 of customers to a higher level of satisfaction)

_____Small (moves about 1/10 of customers to a higher level of satisfaction)

B. What's the second thing you would do to **increase** service performance in this area?

Change	Cost
1.	
2.	
3.	
Total Cost	

How large an effect would you expect this to have on customer satisfaction?

_____Complete (moves about 9/10 of customers to a higher level of satisfaction)

_____Substantial (moves about 2/3 of customers to a higher level of satisfaction)

______Moderate (moves about 1/2 of customers to a higher level of satisfaction)

______Modest (moves about 1/3 of customers to a higher level of satisfaction)

______Small (moves about 1/10 of customers to a higher level of satisfaction)

C. What's the first thing you would do to **economize** on service performance in this area?

Change	**Savings**
1.	
2.	
3.	
Total Savings	

How large an effect would you expect this to have on customer satisfaction?

______Complete (moves about 9/10 of customers to a lower level of satisfaction)

______Substantial (moves about 2/3 of customers to a lower level of satisfaction)

______Moderate (moves about 1/2 of customers to a lower level of satisfaction)

______Modest (moves about 1/3 of customers to a lower level of satisfaction)

______Small (moves about 1/10 of customers to a lower level of satisfaction)

D. What's the second thing you would do to **economize** on service performance in this area?

Change	**Savings**
1.	
2.	
3.	
Total Savings	

How large an effect would you expect this to have on customer satisfaction?

______Complete (moves about 9/10 of customers to a lower level of satisfaction)

______Substantial (moves about 2/3 of customers to a lower level of satisfaction)

______Moderate (moves about 1/2 of customers to a lower level of satisfaction)

______Modest (moves about 1/3 of customers to a lower level of satisfaction)

______Small (moves about 1/10 of customers to a lower level of satisfaction)

5 **Here are four different levels of professionalism you might experience from water utility service representatives when they come to your property to make repairs for non-emergencies. How satisfied would you be with each of the following?** *(Circle one number for each)*

		% Customers reporting	Satisfaction (100 pt. scale)
1.	The service representatives give clear information about what they are doing, listen carefully to what you say, and are polite and friendly	63	98
2.	The service representatives do not give clear information about what they are doing, but listen carefully to what you say, and are polite and friendly	30	59
3.	The service representatives do not give clear information about what they are doing, do not listen carefully to what you say, but are polite and friendly	3	26
4.	The service representatives do not give clear information about what they are doing, do not listen carefully to what you say, and are a little rude or unfriendly	5	5

Reported satisfaction? 83%

Professionalism

A. What's the first thing you would do to **increase** service performance in this area?

Change	**Cost**
1.	
2.	
3.	
Total Cost	

How large an effect would you expect this to have on customer satisfaction?

_____Complete (moves about 9/10 of customers to a higher level of satisfaction)

_____Substantial (moves about 2/3 of customers to a higher level of satisfaction)

_____Moderate (moves about 1/2 of customers to a higher level of satisfaction)

_____Modest (moves about 1/3 of customers to a higher level of satisfaction)

_____Small (moves about 1/10 of customers to a higher level of satisfaction)

B. What's the second thing you would do to **increase** service performance in this area?

Change	**Cost**
1.	
2.	
3.	
Total Cost	

How large an effect would you expect this to have on customer satisfaction?

_____Complete (moves about 9/10 of customers to a higher level of satisfaction)

_____Substantial (moves about 2/3 of customers to a higher level of satisfaction)

_____Moderate (moves about 1/2 of customers to a higher level of satisfaction)

_____Modest (moves about 1/3 of customers to a higher level of satisfaction)

_____Small (moves about 1/10 of customers to a higher level of satisfaction)

C. What's the first thing you would do to **economize** on service performance in this area?

Change	**Savings**
1.	
2.	
3.	
Total Savings	

How large an effect would you expect this to have on customer satisfaction?

_____Complete (moves about 9/10 of customers to a lower level of satisfaction)

_____Substantial (moves about 2/3 of customers to a lower level of satisfaction)

_____Moderate (moves about 1/2 of customers to a lower level of satisfaction)

_____Modest (moves about 1/3 of customers to a lower level of satisfaction)

_____Small (moves about 1/10 of customers to a lower level of satisfaction)

D. What's the second thing you would do to **economize** on service performance in this area?

Change	**Savings**
1.	
2.	
3.	
Total Savings	

How large an effect would you expect this to have on customer satisfaction?

_____Complete (moves about 9/10 of customers to a lower level of satisfaction)

_____Substantial (moves about 2/3 of customers to a lower level of satisfaction)

_____Moderate (moves about 1/2 of customers to a lower level of satisfaction)

_____Modest (moves about 1/3 of customers to a lower level of satisfaction)

_____Small (moves about 1/10 of customers to a lower level of satisfaction)

6 **Here are four different ways water utility service representatives might follow up after they come to your property to make repairs for non-emergencies. How satisfied would you be with each of the following?** *(Circle one number for each)*

		% Customers reporting	Satisfaction (100 pt. scale)
1.	The service representative follows up in writing and with a phone call	41	93
2.	The service representative follows up with a phone call but not in writing	22	84
3.	The service representative follows up in writing but not with a phone call	16	75
4.	The service representative does not follow up	22	19

Reported satisfaction? 78%

Follow-up

A. What's the first thing you would do to **increase** service performance in this area?

Change	**Cost**
1.	
2.	
3.	
Total Cost	

How large an effect would you expect this to have on customer satisfaction?

_____Complete (moves about 9/10 of customers to a higher level of satisfaction)

_____Substantial (moves about 2/3 of customers to a higher level of satisfaction)

_____Moderate (moves about 1/2 of customers to a higher level of satisfaction)

_____Modest (moves about 1/3 of customers to a higher level of satisfaction)

_____Small (moves about 1/10 of customers to a higher level of satisfaction)

B. What's the second thing you would do to **increase** service performance in this area?

Change	**Cost**
1.	
2.	
3.	
Total Cost	

How large an effect would you expect this to have on customer satisfaction?

_____Complete (moves about 9/10 of customers to a higher level of satisfaction)

_____Substantial (moves about 2/3 of customers to a higher level of satisfaction)

_____Moderate (moves about 1/2 of customers to a higher level of satisfaction)

______Modest (moves about 1/3 of customers to a higher level of satisfaction)

______Small (moves about 1/10 of customers to a higher level of satisfaction)

C. What's the first thing you would do to **economize** on service performance in this area?

Change	Savings
1.	
2.	
3.	
Total Savings	

How large an effect would you expect this to have on customer satisfaction?

______Complete (moves about 9/10 of customers to a lower level of satisfaction)

______Substantial (moves about 2/3 of customers to a lower level of satisfaction)

______Moderate (moves about 1/2 of customers to a lower level of satisfaction)

______Modest (moves about 1/3 of customers to a lower level of satisfaction)

______Small (moves about 1/10 of customers to a lower level of satisfaction)

D. What's the second thing you would do to **economize** on service performance in this area?

Change	Savings
1.	
2.	
3.	
Total Savings	

How large an effect would you expect this to have on customer satisfaction?

______Complete (moves about 9/10 of customers to a lower level of satisfaction)

______Substantial (moves about 2/3 of customers to a lower level of satisfaction)

______Moderate (moves about 1/2 of customers to a lower level of satisfaction)

______Modest (moves about 1/3 of customers to a lower level of satisfaction)

______Small (moves about 1/10 of customers to a lower level of satisfaction)

Service Category 3
Customer Service

1 **Here are four different customer service hours a water utility might offer to people who want to speak with a customer service representative either in person or by phone. How satisfied would you be with each of the following?** *(Circle one number for each)*

		% Customers reporting	Satisfaction (100 pt. scale)
1.	Customer service is available 24 hours/day, 7 days/week....	34	94
2.	Customer service is available 12 hours/day (for example, 7 AM to 7 PM), Monday - Friday, and some weekend hours.........................	10	82
3.	Customer service is available 12 hours/day (for example, 7 AM to 7 PM), Monday - Friday..	12	67
4.	Customer service is available 8 hours/day (for example, 9 AM to 5 PM), Monday - Friday..	44	46

Reported satisfaction? 74%

CS Hours of Service

A. What's the first thing you would do to **increase** service performance in this area?

Change	Cost
1.	
2.	
3.	
Total Cost	

How large an effect would you expect this to have on customer satisfaction?

_____Complete (moves about 9/10 of customers to a higher level of satisfaction)

_____Substantial (moves about 2/3 of customers to a higher level of satisfaction)

_____Moderate (moves about 1/2 of customers to a higher level of satisfaction)

_____Modest (moves about 1/3 of customers to a higher level of satisfaction)

_____Small (moves about 1/10 of customers to a higher level of satisfaction)

B. What's the second thing you would do to **increase** service performance in this area?

Change	Cost
1.	
2.	
3.	
Total Cost	

How large an effect would you expect this to have on customer satisfaction?

_____Complete (moves about 9/10 of customers to a higher level of satisfaction)

_____Substantial (moves about 2/3 of customers to a higher level of satisfaction)

______Moderate (moves about 1/2 of customers to a higher level of satisfaction)

______Modest (moves about 1/3 of customers to a higher level of satisfaction)

______Small (moves about 1/10 of customers to a higher level of satisfaction)

C. What's the first thing you would do to **economize** on service performance in this area?

Change	Savings
1.	
2.	
3.	
Total Savings	

How large an effect would you expect this to have on customer satisfaction?

______Complete (moves about 9/10 of customers to a lower level of satisfaction)

______Substantial (moves about 2/3 of customers to a lower level of satisfaction)

______Moderate (moves about 1/2 of customers to a lower level of satisfaction)

______Modest (moves about 1/3 of customers to a lower level of satisfaction)

______Small (moves about 1/10 of customers to a lower level of satisfaction)

D. What's the second thing you would do to **economize** on service performance in this area?

Change	Savings
1.	
2.	
3.	
Total Savings	

How large an effect would you expect this to have on customer satisfaction?

______Complete (moves about 9/10 of customers to a lower level of satisfaction)

______Substantial (moves about 2/3 of customers to a lower level of satisfaction)

______Moderate (moves about 1/2 of customers to a lower level of satisfaction)

______Modest (moves about 1/3 of customers to a lower level of satisfaction)

______Small (moves about 1/10 of customers to a lower level of satisfaction)

2 **Here are three different levels of authority a customer service representative might have about a request for information or action when you call or walk in to your water utility office. How satisfied would you be with each of the following?** *(Circle one number for each)*

		% Customers reporting	Satisfaction (100 pt. scale)
1.	The customer service representative is authorized to make decisions and completes the job easily	62	96
2.	The customer service representative has to seek authorization from others but completes the job easily	32	76
3.	The customer service representative has to seek authorization from others and completes the job with difficulty	6	22

Reported satisfaction? 88%

CS Authorization

A. What's the first thing you would do to **increase** service performance in this area?

Change	**Cost**
1.	
2.	
3.	
Total Cost	

How large an effect would you expect this to have on customer satisfaction?

_____Complete (moves about 9/10 of customers to a higher level of satisfaction)

_____Substantial (moves about 2/3 of customers to a higher level of satisfaction)

_____Moderate (moves about 1/2 of customers to a higher level of satisfaction)

_____Modest (moves about 1/3 of customers to a higher level of satisfaction)

_____Small (moves about 1/10 of customers to a higher level of satisfaction)

B. What's the second thing you would do to **increase** service performance in this area?

Change	**Cost**
1.	
2.	
3.	
Total Cost	

How large an effect would you expect this to have on customer satisfaction?

_____Complete (moves about 9/10 of customers to a higher level of satisfaction)

_____Substantial (moves about 2/3 of customers to a higher level of satisfaction)

_____Moderate (moves about 1/2 of customers to a higher level of satisfaction)

_____Modest (moves about 1/3 of customers to a higher level of satisfaction)

_____Small (moves about 1/10 of customers to a higher level of satisfaction)

C. What's the first thing you would do to **economize** on service performance in this area?

Change	**Savings**
1.	
2.	
3.	
Total Savings	

How large an effect would you expect this to have on customer satisfaction?

_____Complete (moves about 9/10 of customers to a lower level of satisfaction)

_____Substantial (moves about 2/3 of customers to a lower level of satisfaction)

_____Moderate (moves about 1/2 of customers to a lower level of satisfaction)

_____Modest (moves about 1/3 of customers to a lower level of satisfaction)

_____Small (moves about 1/10 of customers to a lower level of satisfaction)

D. What's the second thing you would do to **economize** on service performance in this area?

Change	**Savings**
1.	
2.	
3.	
Total Savings	

How large an effect would you expect this to have on customer satisfaction?

_____Complete (moves about 9/10 of customers to a lower level of satisfaction)

_____Substantial (moves about 2/3 of customers to a lower level of satisfaction)

_____Moderate (moves about 1/2 of customers to a lower level of satisfaction)

_____Modest (moves about 1/3 of customers to a lower level of satisfaction)

_____Small (moves about 1/10 of customers to a lower level of satisfaction)

3 **Here are four different levels of professionalism you might experience from a customer service representative when you call or walk into the office. How satisfied would you be with each of the following?** *(Circle one number for each)*

		% Customers reporting	Satisfaction (100 pt. scale)
1.	The customer service representative is polite and friendly, fully explains utility policy and listens carefully to what you say.	65	98
2.	The customer service representative is polite and friendly, does not explain utility policy but listens carefully to what you say.	31	63
3.	The customer service representative is polite and friendly, does not explain utility policy and does not listen carefully to what you say..	2	26
4.	The customer service representative is a little rude or unfriendly, does not explain utility policy and does not listen carefully to what you say.	2	5

Reported satisfaction? 89%

CS Professionalism

A. What's the first thing you would do to **increase** service performance in this area?

Change	Cost
1.	
2.	
3.	
Total Cost	

How large an effect would you expect this to have on customer satisfaction?

_____Complete (moves about 9/10 of customers to a higher level of satisfaction)

_____Substantial (moves about 2/3 of customers to a higher level of satisfaction)

_____Moderate (moves about 1/2 of customers to a higher level of satisfaction)

_____Modest (moves about 1/3 of customers to a higher level of satisfaction)

_____Small (moves about 1/10 of customers to a higher level of satisfaction)

B. What's the second thing you would do to **increase** service performance in this area?

Change	Cost
1.	
2.	
3.	
Total Cost	

How large an effect would you expect this to have on customer satisfaction?

_____Complete (moves about 9/10 of customers to a higher level of satisfaction)

_____Substantial (moves about 2/3 of customers to a higher level of satisfaction)

_____Moderate (moves about 1/2 of customers to a higher level of satisfaction)

_____Modest (moves about 1/3 of customers to a higher level of satisfaction)

_____Small (moves about 1/10 of customers to a higher level of satisfaction)

C. What's the first thing you would do to **economize** on service performance in this area?

Change	**Savings**
1.	
2.	
3.	
Total Savings	

How large an effect would you expect this to have on customer satisfaction?

_____Complete (moves about 9/10 of customers to a lower level of satisfaction)

_____Substantial (moves about 2/3 of customers to a lower level of satisfaction)

_____Moderate (moves about 1/2 of customers to a lower level of satisfaction)

_____Modest (moves about 1/3 of customers to a lower level of satisfaction)

_____Small (moves about 1/10 of customers to a lower level of satisfaction)

D. What's the second thing you would do to **economize** on service performance in this area?

Change	**Savings**
1.	
2.	
3.	
Total Savings	

How large an effect would you expect this to have on customer satisfaction?

_____Complete (moves about 9/10 of customers to a lower level of satisfaction)

_____Substantial (moves about 2/3 of customers to a lower level of satisfaction)

_____Moderate (moves about 1/2 of customers to a lower level of satisfaction)

_____Modest (moves about 1/3 of customers to a lower level of satisfaction)

_____Small (moves about 1/10 of customers to a lower level of satisfaction)

4 **Here are three experiences you might have on the telephone when you call a customer service representative at your water utility. How satisfied would you be with each of the following?** *(Circle one number for each)*

		% Customers reporting	Satisfaction (100 pt. scale)
1.	An automated system answers the phone. You are able to reach a customer service representative with one input from a short list of options.	41	83
2.	An automated system answers the phone. You are able to reach a customer service representative after three inputs from a long lists of options.	32	38
3.	An automated system answers the phone. You cannot reach a customer service representative.	2	4

Reported satisfaction? 77%

IVR System

A. What's the first thing you would do to **increase** service performance in this area?

Change	Cost
1.	
2.	
3.	
Total Cost	

How large an effect would you expect this to have on customer satisfaction?

_____Complete (moves about 9/10 of customers to a higher level of satisfaction)

_____Substantial (moves about 2/3 of customers to a higher level of satisfaction)

_____Moderate (moves about 1/2 of customers to a higher level of satisfaction)

_____Modest (moves about 1/3 of customers to a higher level of satisfaction)

_____Small (moves about 1/10 of customers to a higher level of satisfaction)

B. What's the second thing you would do to **increase** service performance in this area?

Change	Cost
1.	
2.	
3.	
Total Cost	

How large an effect would you expect this to have on customer satisfaction?

_____Complete (moves about 9/10 of customers to a higher level of satisfaction)

_____Substantial (moves about 2/3 of customers to a higher level of satisfaction)

_____Moderate (moves about 1/2 of customers to a higher level of satisfaction)

_____Modest (moves about 1/3 of customers to a higher level of satisfaction)

_____Small (moves about 1/10 of customers to a higher level of satisfaction)

C. What's the first thing you would do to **economize** on service performance in this area?

Change	**Savings**
1.	
2.	
3.	
Total Savings	

How large an effect would you expect this to have on customer satisfaction?

_____Complete (moves about 9/10 of customers to a lower level of satisfaction)

_____Substantial (moves about 2/3 of customers to a lower level of satisfaction)

_____Moderate (moves about 1/2 of customers to a lower level of satisfaction)

_____Modest (moves about 1/3 of customers to a lower level of satisfaction)

_____Small (moves about 1/10 of customers to a lower level of satisfaction)

D. What's the second thing you would do to **economize** on service performance in this area?

Change	**Savings**
1.	
2.	
3.	
Total Savings	

How large an effect would you expect this to have on customer satisfaction?

_____Complete (moves about 9/10 of customers to a lower level of satisfaction)

_____Substantial (moves about 2/3 of customers to a lower level of satisfaction)

_____Moderate (moves about 1/2 of customers to a lower level of satisfaction)

_____Modest (moves about 1/3 of customers to a lower level of satisfaction)

_____Small (moves about 1/10 of customers to a lower level of satisfaction)

5 **Here are four different lengths of time you might have to wait on hold before speaking to a customer service representative when you telephone the water utility office. How satisfied would you be with each of the following?** *(Circle one number for each)*

		% Customers reporting	Satisfaction (100 pt. scale)
1.	You do not have to wait on hold. You speak to a customer service representative right away..	31	96
2.	You are on hold for 20 seconds before speaking to a customer service representative	27	84
3.	You are on hold for 45 seconds before speaking to a customer service representative	25	69
4.	You are on hold for 120 seconds (2 minutes) before speaking to a customer service representative...	17	33

Reported satisfaction? 81%

Hold Time (Call-in)

A. What's the first thing you would do to **increase** service performance in this area?

Change	Cost
1.	
2.	
3.	
Total Cost	

How large an effect would you expect this to have on customer satisfaction?

_____Complete (moves about 9/10 of customers to a higher level of satisfaction)

_____Substantial (moves about 2/3 of customers to a higher level of satisfaction)

_____Moderate (moves about 1/2 of customers to a higher level of satisfaction)

_____Modest (moves about 1/3 of customers to a higher level of satisfaction)

_____Small (moves about 1/10 of customers to a higher level of satisfaction)

B. What's the second thing you would do to **increase** service performance in this area?

Change	Cost
1.	
2.	
3.	
Total Cost	

How large an effect would you expect this to have on customer satisfaction?

_____Complete (moves about 9/10 of customers to a higher level of satisfaction)

_____Substantial (moves about 2/3 of customers to a higher level of satisfaction)

_____Moderate (moves about 1/2 of customers to a higher level of satisfaction)

_____Modest (moves about 1/3 of customers to a higher level of satisfaction)

_____Small (moves about 1/10 of customers to a higher level of satisfaction)

C. What's the first thing you would do to **economize** on service performance in this area?

Change	**Savings**
1.	
2.	
3.	
Total Savings	

How large an effect would you expect this to have on customer satisfaction?

_____Complete (moves about 9/10 of customers to a lower level of satisfaction)

_____Substantial (moves about 2/3 of customers to a lower level of satisfaction)

_____Moderate (moves about 1/2 of customers to a lower level of satisfaction)

_____Modest (moves about 1/3 of customers to a lower level of satisfaction)

_____Small (moves about 1/10 of customers to a lower level of satisfaction)

D. What's the second thing you would do to **economize** on service performance in this area?

Change	**Savings**
1.	
2.	
3.	
Total Savings	

How large an effect would you expect this to have on customer satisfaction?

_____Complete (moves about 9/10 of customers to a lower level of satisfaction)

_____Substantial (moves about 2/3 of customers to a lower level of satisfaction)

_____Moderate (moves about 1/2 of customers to a lower level of satisfaction)

_____Modest (moves about 1/3 of customers to a lower level of satisfaction)

_____Small (moves about 1/10 of customers to a lower level of satisfaction)

6 **Here are four different lengths of time you might have to wait in line before being served by a customer service representative when you walk in to the water utility office. How satisfied would you be with each of the following?** *(Circle one number for each)*

		% Customers reporting	Satisfaction (100 pt. scale)
1.	You do not have to wait in line. A customer service representative serves you right away.	33	97
2.	You have to wait in line for 1 minute before a customer service representative serves you	33	88
3.	You have to wait in line for 5 minutes before a customer service representative serves you.	24	56
4.	You have to wait in line for 10 minutes before a customer service representative serves you	9	27

Reported satisfaction? 78%

Wait Time (Walk-in)

A. What's the first thing you would do to **increase** service performance in this area?

Change	Cost
1.	
2.	
3.	
Total Cost	

How large an effect would you expect this to have on customer satisfaction?

_____Complete (moves about 9/10 of customers to a higher level of satisfaction)

_____Substantial (moves about 2/3 of customers to a higher level of satisfaction)

_____Moderate (moves about 1/2 of customers to a higher level of satisfaction)

_____Modest (moves about 1/3 of customers to a higher level of satisfaction)

_____Small (moves about 1/10 of customers to a higher level of satisfaction)

B. What's the second thing you would do to **increase** service performance in this area?

Change	Cost
1.	
2.	
3.	
Total Cost	

How large an effect would you expect this to have on customer satisfaction?

_____Complete (moves about 9/10 of customers to a higher level of satisfaction)

_____Substantial (moves about 2/3 of customers to a higher level of satisfaction)

_____Moderate (moves about 1/2 of customers to a higher level of satisfaction)

_____Modest (moves about 1/3 of customers to a higher level of satisfaction)

_____Small (moves about 1/10 of customers to a higher level of satisfaction)

C. What's the first thing you would do to **economize** on service performance in this area?

Change	Savings
1.	
2.	
3.	
Total Savings	

How large an effect would you expect this to have on customer satisfaction?

_____Complete (moves about 9/10 of customers to a lower level of satisfaction)

_____Substantial (moves about 2/3 of customers to a lower level of satisfaction)

_____Moderate (moves about 1/2 of customers to a lower level of satisfaction)

_____Modest (moves about 1/3 of customers to a lower level of satisfaction)

_____Small (moves about 1/10 of customers to a lower level of satisfaction)

D. What's the second thing you would do to **economize** on service performance in this area?

Change	Savings
1.	
2.	
3.	
Total Savings	

How large an effect would you expect this to have on customer satisfaction?

_____Complete (moves about 9/10 of customers to a lower level of satisfaction)

_____Substantial (moves about 2/3 of customers to a lower level of satisfaction)

_____Moderate (moves about 1/2 of customers to a lower level of satisfaction)

_____Modest (moves about 1/3 of customers to a lower level of satisfaction)

_____Small (moves about 1/10 of customers to a lower level of satisfaction)

Service Category 4
Customer Communications

1 **Here are four descriptions of how often a water utility might communicate with the public about its operations and about issues of special interest to its customers. How satisfied would you be with each of the following?** *(Circle one number for each)*

	% Customers reporting	Satisfaction (100 pt. scale)
1. The utility communicates monthly	82	87
2. The utility communicates 4 times/year	14	71
3. The utility communicates once/year	0	38
4. The utility communicates only in emergencies	4	17

Reported satisfaction? 93%

Communication Frequency

A. What's the first thing you would do to **increase** service performance in this area?

Change	Cost
1.	
2.	
3.	
Total Cost	

How large an effect would you expect this to have on customer satisfaction?

_____Complete (moves about 9/10 of customers to a higher level of satisfaction)

_____Substantial (moves about 2/3 of customers to a higher level of satisfaction)

_____Moderate (moves about 1/2 of customers to a higher level of satisfaction)

_____Modest (moves about 1/3 of customers to a higher level of satisfaction)

_____Small (moves about 1/10 of customers to a higher level of satisfaction)

B. What's the second thing you would do to **increase** service performance in this area?

Change	Cost
1.	
2.	
3.	
Total Cost	

How large an effect would you expect this to have on customer satisfaction?

_____Complete (moves about 9/10 of customers to a higher level of satisfaction)

_____Substantial (moves about 2/3 of customers to a higher level of satisfaction)

______Moderate (moves about 1/2 of customers to a higher level of satisfaction)

______Modest (moves about 1/3 of customers to a higher level of satisfaction)

______Small (moves about 1/10 of customers to a higher level of satisfaction)

C. What's the first thing you would do to **economize** on service performance in this area?

Change	**Savings**
1.	
2.	
3.	
Total Savings	

How large an effect would you expect this to have on customer satisfaction?

______Complete (moves about 9/10 of customers to a lower level of satisfaction)

______Substantial (moves about 2/3 of customers to a lower level of satisfaction)

______Moderate (moves about 1/2 of customers to a lower level of satisfaction)

______Modest (moves about 1/3 of customers to a lower level of satisfaction)

______Small (moves about 1/10 of customers to a lower level of satisfaction)

D. What's the second thing you would do to **economize** on service performance in this area?

Change	**Savings**
1.	
2.	
3.	
Total Savings	

How large an effect would you expect this to have on customer satisfaction?

______Complete (moves about 9/10 of customers to a lower level of satisfaction)

______Substantial (moves about 2/3 of customers to a lower level of satisfaction)

______Moderate (moves about 1/2 of customers to a lower level of satisfaction)

______Modest (moves about 1/3 of customers to a lower level of satisfaction)

______Small (moves about 1/10 of customers to a lower level of satisfaction)

2 **Here are four schedules on which a water utility might distribute routine (non-emergency) information about water quality. How satisfied would you be with each of the following?** *(Circle one number for each)*

		% Customers reporting	Satisfaction (100 pt. scale)
1.	Water quality information is distributed in the mail monthly	58	86
2.	Water quality information is distributed in the mail every three months	19	73
3.	Water quality information is distributed in the mail once a year	19	36
4.	Water quality information is available on the internet	3	36

Reported satisfaction? 86%

Water Quality Information

A. What's the first thing you would do to **increase** service performance in this area?

Change	**Cost**
1.	
2.	
3.	
Total Cost	

How large an effect would you expect this to have on customer satisfaction?

_____Complete (moves about 9/10 of customers to a higher level of satisfaction)

_____Substantial (moves about 2/3 of customers to a higher level of satisfaction)

_____Moderate (moves about 1/2 of customers to a higher level of satisfaction)

_____Modest (moves about 1/3 of customers to a higher level of satisfaction)

_____Small (moves about 1/10 of customers to a higher level of satisfaction)

B. What's the second thing you would do to **increase** service performance in this area?

Change	**Cost**
1.	
2.	
3.	
Total Cost	

How large an effect would you expect this to have on customer satisfaction?

_____Complete (moves about 9/10 of customers to a higher level of satisfaction)

_____Substantial (moves about 2/3 of customers to a higher level of satisfaction)

______Moderate (moves about 1/2 of customers to a higher level of satisfaction)

______Modest (moves about 1/3 of customers to a higher level of satisfaction)

______Small (moves about 1/10 of customers to a higher level of satisfaction)

C. What's the first thing you would do to **economize** on service performance in this area?

Change	**Savings**
1.	
2.	
3.	
Total Savings	

How large an effect would you expect this to have on customer satisfaction?

______Complete (moves about 9/10 of customers to a lower level of satisfaction)

______Substantial (moves about 2/3 of customers to a lower level of satisfaction)

______Moderate (moves about 1/2 of customers to a lower level of satisfaction)

______Modest (moves about 1/3 of customers to a lower level of satisfaction)

______Small (moves about 1/10 of customers to a lower level of satisfaction)

D. What's the second thing you would do to **economize** on service performance in this area?

Change	**Savings**
1.	
2.	
3.	
Total Savings	

How large an effect would you expect this to have on customer satisfaction?

______Complete (moves about 9/10 of customers to a lower level of satisfaction)

______Substantial (moves about 2/3 of customers to a lower level of satisfaction)

______Moderate (moves about 1/2 of customers to a lower level of satisfaction)

______Modest (moves about 1/3 of customers to a lower level of satisfaction)

______Small (moves about 1/10 of customers to a lower level of satisfaction)

3 **Here are four different levels of information from your water utility you might be able to get on the internet. How satisfied would you be with each of the following?** *(Circle one number for each)*

		% Customers reporting	Satisfaction (100 pt. scale)
1.	Individual account information, bill payment history and general information	36	66
2.	Individual account information and general information	6	60
3.	Only general information	36	45
4.	No internet information	21	30

Reported satisfaction? 75%

Internet Access

A. What's the first thing you would do to **increase** service performance in this area?

Change	Cost
1.	
2.	
3.	
Total Cost	

How large an effect would you expect this to have on customer satisfaction?

_____Complete (moves about 9/10 of customers to a higher level of satisfaction)

_____Substantial (moves about 2/3 of customers to a higher level of satisfaction)

_____Moderate (moves about 1/2 of customers to a higher level of satisfaction)

_____Modest (moves about 1/3 of customers to a higher level of satisfaction)

_____Small (moves about 1/10 of customers to a higher level of satisfaction)

B. What's the second thing you would do to **increase** service performance in this area?

Change	Cost
1.	
2.	
3.	
Total Cost	

How large an effect would you expect this to have on customer satisfaction?

_____Complete (moves about 9/10 of customers to a higher level of satisfaction)

_____Substantial (moves about 2/3 of customers to a higher level of satisfaction)

______Moderate (moves about 1/2 of customers to a higher level of satisfaction)

______Modest (moves about 1/3 of customers to a higher level of satisfaction)

______Small (moves about 1/10 of customers to a higher level of satisfaction)

C. What's the first thing you would do to **economize** on service performance in this area?

Change	**Savings**
1.	
2.	
3.	
Total Savings	

How large an effect would you expect this to have on customer satisfaction?

______Complete (moves about 9/10 of customers to a lower level of satisfaction)

______Substantial (moves about 2/3 of customers to a lower level of satisfaction)

______Moderate (moves about 1/2 of customers to a lower level of satisfaction)

______Modest (moves about 1/3 of customers to a lower level of satisfaction)

______Small (moves about 1/10 of customers to a lower level of satisfaction)

D. What's the second thing you would do to **economize** on service performance in this area?

Change	**Savings**
1.	
2.	
3.	
Total Savings	

How large an effect would you expect this to have on customer satisfaction?

______Complete (moves about 9/10 of customers to a lower level of satisfaction)

______Substantial (moves about 2/3 of customers to a lower level of satisfaction)

______Moderate (moves about 1/2 of customers to a lower level of satisfaction)

______Modest (moves about 1/3 of customers to a lower level of satisfaction)

______Small (moves about 1/10 of customers to a lower level of satisfaction)

REFERENCES

Baumgartner, R.M. and T.A. Heberlein. 1984. Recent Research on Mailed Questionnaire Response Rate. In *New Directions in Program Evaluation: Making Effective Use of Mailed Questionnaires*. Edited by D. C. Lockhart. New York: Jossey-Bass Publishers.

Chvatal, V. 1983. Linear Programming. New York: W.H. Freeman.

Colbourne, J., A. Owen, L. Godfrey, J. Brisco, D. Holt. 2002. Tools and Methods to Effectively Measure Customer Perceptions. Denver, CO: AwwaRF.

Dillman, D.A. 2000. *Mail and Internet Surveys: The Tailored Design Method,* Second Edition. New York: J Wiley.

Gass, S. 1970. An Illustrated Guide to Linear Programming. New York: Dover.

Goodman, P. S., M. Fichman, F. J. Lerch, and P. R. Snyder. 1995. Customer-Firm Relationships, Involvement, and Customer Relations. *Academy of Management Journal,* 38(5): 1310-1324.

Hayes, B.E. 1998. *Measuring Customer Satisfaction: Survey Design, Use, and Statistical Analysis Methods*, second edition. Milwaukee, WI: ASQ Quality Press.

Heberlein, T.A. and R.M. Baumgartner. 1978. Factors Affecting Response Rates to Mailed Questionnaires: A Quantitative Analysis of the Published Literature. *American Sociological Review*, 43(4): 447-462.

Krueger, R.A. and M.A. Casey. 2000. *Focus Groups: A Practical Guide for Applied Research.* Thousand Oaks, CA: Sage Publications.

Morgan, D.L. 1997. *Focus Groups as Qualitative Research.* Thousand Oaks, CA: Sage Publications.

Nykamp, M. 2001. *The Customer Differential: The Complete Guide to Implementing Customer Relationship Management*. New York: AMACOM.

Oliver, R. L. 1980. A Cognitive Model of the Antecedents and Consequences of Satisfaction Decisions. *Journal of Marketing Research*, 17(November): 460-469.

Oliver, R. L. 1997. *Satisfaction: A Behavioral Perspective on the Consumer.* New York: McGraw Hill.

Olstein, M.A., M.J. Stanford, and C.E. Day. 2001. *Best Practices for a Continually Improving Customer Responsive Organization.* Denver, CO: AwwaRF.

Rust, R.T., V.A. Zeithaml, and, K.N. Lemon. 2000. *Driving Customer Equity: How Customer Lifetime Value is Reshaping Corporate Strategy.* New York: Free Press.

Sharma, S, R. W. Niedrich, and G. Dobbins. 1999. A Framework for Monitoring Customer Satisfaction. *Industrial Marketing Management*, 28: 231-243.

Vavra, T.G. 1995. Aftermarketing: How to Keep Customers for Life Through Relationship Marketing. Chicago: Irwin.

Wiersema, F. 1996. *Customer Intimacy: Pick Your Partners, Shape Your Culture, Win Together.* Santa Monica, CA: Knowledge Exchange.

ABBREVIATIONS

Ariz.	Arizona
AWWA	American Water Works Association
AwwaRF	Awwa Research Foundation
D.C.	District of Columbia
D.C. WASA	District of Columbia Water and Sewer Authority
Fla.	Florida
PAC	Project advisory committee
Tenn.	Tennessee
U.S.	United States
U.S.A.	United States of America

IWA Publishing's authorised EU representative for General Product Safety Regulations is Diane D'Arras, 15 rue Duret, 75116 Paris, France, e-mail: safety@iwap.co.uk.

Printed and bound by CPI Group (UK) Ltd, Croydon, CR0 4YY
19/09/2026
02219351-0004